Structures of agricultural education

Titles in this series:

Elements of the structure of agricultural education in the United States of America

David L. Howell
William I. Lindley
Raymond H. Morton
Glenn Z. Stevens
Edgar Paul Yoder

Unesco

The authors are responsible for the choice and the presentation of the facts contained in this book and for the opinions expressed therein, which are not necessarily those of Unesco and do not commit the Organization.

Published by the United Nations Educational,
Scientific and Cultural Organization
7 place de Fontenoy, 75700 Paris
Printed by Imprimerie Darantiere, Quetigny

ISBN 92-3-102056-0

Preface

Human resources are recognized as being the key factor in social and economic progress, and their promotion and development are considered as having top priority, particularly for rural areas, and above all in developing countries. Thus, efforts to improve the qualitative and quantitative standards of agricultural education and training are essential for progress in the rural sector, for increasing its productivity and for ensuring the harmonious development and well-being of rural communities.

The World Conference on Agricultural Education and Training held in Copenhagen (1970) and sponsored jointly by FAO, ILO and Unesco stressed the need for more information and studies related to the problems of education and rural development. The dissemination and exchange of information on existing structures of agricultural education are considered vital to the improvement of its quality.

Recognizing this need, Unesco has published a series of detailed country studies on the 'Structures of Agricultural Education', the aim of which is to provide Member States with precise information on the subject, and to facilitate comparisons of structure and terminology. A comparative study will be prepared as a synthesis of the individual monographs in this series.

The present study, devoted to agricultural education in the United States of America, gives special attention to its integration in general education, its place in continuing education for adults, the various levels of formal agricultural education and the co-operative extension service. It also describes a curriculum development model for vocational agriculture.

The study was prepared by David L. Howell (B.S., M.E., Ph.D.; Assistant Professor, Department of Agricultural Education), William I. Lindley (Ph.D.; Assistant Professor, Agricultural Education and Extension Education), Raymond H. Morton (B.Sc., Ph.D.; Assistant Professor of Agricultural Education), Glenn Z. Stevens (Ph.D.; Professor Emeritus, Department of Agricultural Education), and Edgar Paul Yoder (Ph.D.; Assistant Professor, Agricultural Education), all of the Pennsylvania State University, under the auspices of the United States National Commission for Unesco. Unesco wishes to thank the authors and the National Commission for having undertaken the considerable work involved.

Contents

Foreword

Agricultural and extension education continue to perform a vital role in the communication of new technology for those engaged in production agriculture and agribusiness. They serve as an important link between the researcher who develops the new technology and those engaged in occupations that can benefit from it. The value of a new technology depends on its proper introduction and adoption.

This book identifies programmes that prepare secondary and post-secondary school students for employment in agricultural occupations and provide adults with the means to introduce new technology into production agriculture and agribusiness occupations. Because of the broad scope of this book, a bibliography is included with each chapter.

Producing food and providing for its storage and equitable distribution have always been major problems of the world. Education is an ingredient in the solution to these problems. Much is being done in the development of new agricultural technology. However, many areas of the world lack an effective system of making this technology available for use by all producers of agricultural products and all agribusinesses in an understandable and acceptable way. Governments and educators must encourage greater development of agricultural and extension education programmes as a part of their total development programme for increases in the available food supply to be realized by all.

David L. Howell
Pennsylvania State University

Historical perspective of agricultural education

Glenn Z. Stevens and David L. Howell

The art of agriculture is becoming more sophisticated with each succeeding generation of farmers. Technological developments follow discoveries in the biological and physical sciences and often stimulate further basic research. Labour-saving mechanization in farming and in the manufacturing of agricultural supplies and products is a development of agricultural engineering.

National and international growth and change make the understanding of the principles and interpretations of political science important to everyone involved in policy determination and in planning. Management and marketing of agricultural products is big business. Individual career guidance, preparation for family and community living and development of essential communication skills are needed to equip youth and adults for careers in agricultural occupations. To bring these new developments in agriculture to those preparing for, or engaged in, an agricultural occupation requires teachers who are knowledgeable and up to date in all areas of agriculture. The teaching of vocational agriculture and the co-operative extension service are ways of bringing these new developments into the secondary-school classrooms and to the young and adult farmers in the area.

Early settlers in North America brought with them the practices in crop production and animal husbandry generally used in Europe. They imported seeds, plants, breeding stock and hand tools. The struggle for survival at a subsistence level of farming was the primary way of life in the United States when the nation was formed. Commercial agriculture gradually emerged as eastern cities grew, water transportation was improved, and railroads were built. These changes occurred between 1800 and 1850, a period in which the invention of horse-drawn tillage and harvesting implements reduced manual labour in farming operations. Knowledge, however, continued to be passed from fathers to sons verbally, and was meagre in quantity. Books and magazines were few, many adults were illiterate, and schools enrolled children for only a few years.

It has been traditional to consider that soil, particularly fertile topsoil, is the basic resource for agriculture. Sufficient food production depends upon adequate cropland areas, water, climate, soil fertility and the use of current technology. Education makes possible the efficient use of all forms of power, mechanization, science, and technology in modifying the physical, chemical, biological, economic and social aspects of the total environment.

Leaders in the United States, notably legislators and academically educated professionals who organized state and local agricultural development, tried on many occasions in the first seventy years of the nation's history to obtain Federal support for education, experimentation, and dissemination of information relating to agriculture. Success came in 1862 when separate Acts of Congress created the United States Department of Agriculture and made grants of public land to each state. The sale of this land provided funds to establish a college of agriculture and mechanic arts. The Department of Agriculture was raised to Cabinet level in 1889. The Land-Grant College

(Morrill) Act, administered by the Department, resulted in the creation of residential agricultural education institutions and in the eventual establishment of experimental stations and co-operative extension work.

The First Morrill Act of 1862 accomplished its purpose and is considered to be the origin of the present comprehensive system of public education in agriculture in the United States. The Act of 1890 for the Further Endowment of Land-Grant Colleges (Second Morrill Act) made additional funds available to each state and territory and required that in states that had established separate colleges for white and coloured students there should be a just and equitable division of the benefits of the Act.

Three other Acts of Congress became basic integral parts of the national programme in agricultural education. The Hatch Act of 1887 authorized appropriations to the states and territories for support of agricultural experimental stations. Since 1888, the Hatch fund has been included in the annual appropriation for the Department of Agriculture. The Act of 1914 Providing for Cooperative Extension Work (Smith-Lever Act) made permanent annual appropriations to facilitate the relationship between the land-grant college or colleges in each state and the Department of Agriculture. This resulted in the development of programmes for the diffusion of useful and practical information on subjects relating to agriculture and home economics. The National Vocational Education (Smith-Hughes) Act, passed in 1917, provided permanent annual appropriations for the promotion of vocational education; co-operation with the states in the payment of salaries of teachers, supervisors, or directors of agricultural subjects and teachers of trade, home economics and industrial subjects; and co-operation with the states in the preparation of teachers of vocational subjects.

Scope of agricultural education by types of agriculture-related industries

Production agriculture

The scope of agricultural education can be described in a functional way by listing the types of industries that employ persons needing knowledge and skills taught in organized agricultural programmes. Production is the oldest and largest area of the total agriculture and agribusiness industry.

Data in Table 1 from the 1974 Census of Agriculture show that in the United States 1,695,047 farms with sales of $2,500 and over were nearly equally divided between crop production and livestock production. The long-term trends have been towards larger farms with higher average value of land and buildings. Individual owner-operators manage most of the farms and ranches; included are family businesses and partnerships. Tenants operate 11.3 per cent of all farms.

The Department of Commerce, Bureau of the Census, changed the definition of a farm to include all land on which agricultural operations were conducted at any time in the census year under the day-to-day control of an individual management, and from which $1,000 or more of agricultural products were sold during the census year. Control may have been exercised through ownership or management, or through a lease, rental, or cropping arrangement. Places having less than the minimum $1,000 sales in the census year were also counted as farms if they could normally be expected to produce agricultural products in sufficient quantity to meet the requirements of the definition. There were 616,728 farms in this category. Two-thirds of them were operated by persons whose principal occupation was other than farming.

TABLE 1. Number, size and value of farms with sales put at $2,500 and over by the Standard Industrial Classification of Establishments

	Number of farms	Average hectares per farm	Average value land and bldgs in dollars
Cash grain farms	580 254	196	206 661
Cotton farms	30 725	234	247 068
Tobacco farms	95 493	52	64 507
Sugar crop, Irish potato, hay, peanut and other field-crop farms	81 415	193	209 115
Vegetable and melon farms	19 548	97	235 233
Fruit and tree nut farms	51 270	60	212 474
Horticultural specialty farms	19 678	25	124 398
General farms, primarily crop	44 659	212	205 951
Livestock farms, except dairy, poultry and animal specialty	493 816	363	189 890
Dairy farms	196 057	112	136 954
Poultry and egg farms	42 690	52	96 846
Animal specialty farms	11 167	74	125 990
General farms, principally livestock	14 995	162	160 357
Farms not classified by SIC	13 280	261	187 684
TOTAL	1 695 047	216	182 231

Source : Census of Agriculture 1974, U.S. Data.

Agribusiness

A fast-growing array of agricultural services performed by agribusiness establishments is outlined in Table 2. There were 61,347 establishments with 500,000 paid workers, mainly part-time technical and skilled employees, representing a $1,206 million annual payroll. Included were soil-preparation services, crop services, veterinary services, animal services other than veterinary, farm labour and management services, and landscape and horticultural services.

Many of these occupations require employees with agricultural production backgrounds. As smaller numbers of the population are working in production agriculture it is becoming more difficult for agribusiness to locate the employees needed.

Agricultural supplies and services

Manufacturing and processing industries provide an expanding variety of supplies and equipment needed in production agriculture. Chemicals, livestock feeds, crop seeds, and fertilizers are some of the essential inputs to farming. Power, machinery, buildings, and mechanical equipment are supplied by specialized manufacturers, distributors, and retailers.

Agricultural products processing and marketing

Agricultural products processing and marketing functions may begin on the farm as preliminary stages of assembling, grading, temperature control, drying, storing and

TABLE 2. Establishments and number of employees in agricultural services in the United States in 1974

| | | Paid employees | |
	Establishments	150 days or more	Less than 150 days
Total number of services	61 347	159 616	341 052
Soil preparation services	730	1 425	1 832
Crop services	5 140	20 380	88 191
Planting, cultivating, and protection	1 392	5 512	6 865
Harvesting, primarily by machine	1 288	2 153	7 543
Preparation for market	934	8 575	44 596
Cotton ginning	1 522	4 138	29 183
General crop services	4	2	4
Veterinary services	10 452	32 126	20 198
Cattle, hogs, sheep, goats, and poultry	3 081	5 985	4 221
Dogs, cats, horses, bees, fish, rabbits, other fur-bearing animals, birds (except poultry), and other pets	7 371	26 141	15 977
Animal services, except veterinary	10 502	13 651	15 591
Cattle, hogs, sheep, goats, and poultry	1 344	4 725	3 374
Dogs, cats, horses, bees, fish, rabbits, other fur-bearing animals, birds (except poultry), and other pets	9 158	8 926	12 217
Farm labour and management services	905	32 064	129 299
Farm labour contractors and crew leaders	504	21 939	101 049
Farm management services, complete maintenance and management	401	10 125	28 250
Landscape and horticultural services	33 618	59 970	85 941
Landscape counseling and planning	1 560	3 717	3 685
Lawn and garden services	24 940	25 100	46 940
Ornamental shrub and tree services	7 118	31 153	35 316

Source : 1974 Census of Agriculture.

transporting. But modern agribusiness establishments now move food and other agricultural products most efficiently into the hands of consumers. Management and operation of marketing co-operatives and of corporate food-manufacturing businesses employ more workers than the commercial farms of the nation.

Renewable natural resources

Environmental resource management is a part of agricultural education in that it contributes to the conservation, protection and regulation, and recreational utilization of renewable natural resources. Land, water and air can be well managed by persons with a knowledge of basic ecological factors. Education in agriculture also equips graduates to make decisions and carry out practices favourable to crop and livestock productivity, to environmental pollution control, as well as to the best uses of open space, forests and wildlife habitats. Outcomes may be measured ultimately in economic returns, in human health, and in aesthetic satisfaction.

Communications

Agricultural communications is another area of employment which young men and women find attractive and rewarding. In agribusiness firms, the career opportunities are in advertising, public relations and supporting services that transmit and disseminate information among departments and individuals in the business and its hierarchy of integrated trade organizations. Television and radio, journalism, and photography need a continuous supply of new and replacement employees with all levels of expertise. Agricultural editors and publications staff in university and research institutions, on biological science journals and farm magazines, and in multi-media news and instructional materials production are usually graduates of formal programmes in communications.

Education

Education is, in itself, a professional field, and agricultural education is a distinct function of the land-grant universities. A compelling demand for more knowledge developed in the United States between 1865 and 1885 and brought about the establishment of agricultural experimental stations. Experiments with lectures, exhibits and institutes to encourage increased efficiency were well received by the small numbers of farmers who could be reached in different states. Resident college enrolment increased in departments where superior instruction motivated the learners.

Qualified teachers of agriculture are needed in secondary schools and area vocational schools, in two-year post-secondary technical institutes and community colleges, and in baccalaureate, master's, and doctoral degree programmes in universities. Each state has standards and requirements that must be met by baccalaureate graduates who apply for a teaching certificate entitling the holder to teach agriculture in secondary schools. Issuing of certificates is a function of the state department of education, but responsibility for approving candidates is often delegated to accredited university teacher education faculties. A certificate, or credential, to teach any area or subject in agriculture is not needed in post-secondary institutions, in colleges or universities, or in the co-operative extension service. Each type of institution maintains and makes efforts to improve a programme of instructor upgrading in teaching methodology and learning effectiveness. Government service or employment with foundations in positions needing knowledge in agriculture and skill in teaching has, in some universities, brought extension education, adult education and international education in agriculture to the status of a department in which students can major.

Research and development

Research and development are graduate-school activities of the land-grant universities. Working together, graduate students and faculty members systematically use standard techniques and innovative procedures as they strive towards identifying solutions to problems. From the beginning of Federal and state support, experimental station research has been on a project unit basis with prior approval of a formal written proposal and sustaining supervision.

Research in the agricultural industry is usually product-oriented, but research workers are likely to be identified by their specialization in an area of scientific

15

competence, such as biology, chemistry, economics, food technology, genetics, animal nutrition, plant pathology, etc. Research breakthroughs create new jobs, stimulate the setting up of new establishments and expand the capacity, efficiency and profitability of commercial farms. A hallmark of an outstanding undergraduate curriculum is that it includes some reasonably meaningful experiences in learning through applied research.

Structure of agricultural education by institutions and organizations

The structure of agricultural education in the United States is presented in this book in five parts. Each is treated in a separate chapter. As the chapter titles indicate, the structure is by type of institutions. There are varying degrees of interdependence among them, with mutually beneficial outcomes, but each institution or organization is primarily an autonomous unit.

The second chapter describes agricultural education in public schools. In general, these schools serve small geographic areas and are under local administrative control in the political subdivisions of each state. The third chapter is devoted to programmes of two-year post-secondary education in agriculture in technical schools and community colleges. Most of the offerings at post-secondary level have been initiated in recent years. In the fourth chapter, which covers university education in agriculture, the major emphasis is on teacher-preparation programmes for those in agriculture and extension education occupations. The fifth chapter, on the co-operative extension service, outlines several components of the total structure that provide continuing informal and non-formal education to citizens throughout their lives. Orientation to family and community economic and social concerns, responsible stewardship of renewable natural resources, and efficient public investment in agricultural research are well-known and favourably regarded national objectives. An important part of an educational programme is the development and distribution of teaching materials. The sixth chapter, on curriculum development, instructional materials and instructional services, identifies the process by which such materials are developed and disseminated. These materials are essential to support teaching. Examples of sources for such materials are provided. Universities do much to meet the needs of teachers but private publishers and agribusiness firms also provide useful curriculum materials.

Land-grant universities continue to be the principal sources of teachers for all types of organized programmes of agricultural education. Faculty members who teach in vocational education in public secondary and post-secondary institutions in all states are required to have taken prescribed preservice professional courses in education, including student teaching. Throughout the total structure of agricultural education, the persons employed to teach, counsel, manage, supervise, and communicate information and services are of baccalaureate level or hold higher degrees. Their selection includes careful attention to evidence of potential success in dealing with people.

Integration of agricultural education and general education

In the early 1970s much emphasis was given to the fact that the educational system in the United States should be oriented to preparing all students for their chosen occupations. Such an education would not end at a given point in time for an individual, but would be a continuing process throughout the life of the person. This approach became known as career education or life-long education. An example of a career-education model is given in Figure 1. The model shows how vocational education, of which vocational agriculture is a part, relates to the total educational system. Before models such as this were implemented, schools offered many courses that did not include preparation for specific occupations. Students not planning to continue their education beyond secondary school were enrolled in classes that had little meaning for them. As a result of the career-education model, focus is now given in general education courses to training for occupations of specific interest to each student. For example, reading assignments in English classes are related to the student's occupational interests. The problems presented in mathematics classes involve computational procedures needed in the student's occupational choice. In this way, general education helps to provide the background needed for vocational classes, in entry-level occupations and for career advancement.

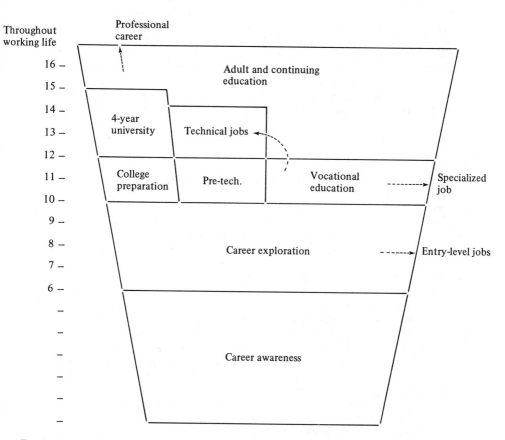

Fig. 1. Career-education model, school-based example.

Agricultural education in public schools[1]

Glenn Z. Stevens and David L. Howell

Vocational-agriculture programmes emphasize the preparation of secondary-school students for gainful employment in production agriculture and in businesses related to agriculture. Practical experience in the school agricultural facilities and on a farm or in an agribusiness are key factors in successful educational programmes. Future Farmers of America (FFA) provides leadership training for the students enrolled in these agriculture programmes and organizes contests with awards for outstanding students. The students are visited by their vocational-agriculture instructor on their farm or in the agribusiness where they work. The vocational-agriculture curriculum is designed to meet each student's occupational needs for employment and is an important part of maintaining a relevant educational programme.

Instruction in agriculture is offered in many primary schools in order to provide students with an understanding of how the food they eat is produced. Adults in many communities are also offered instruction so as to inform them of new developments in agriculture and in businesses related to agriculture.

Secondary-level agricultural education

Vocational education in agriculture is an elective area in the secondary-school curriculum. Depending on the needs of the community and on facilities and instructor availability, courses may be offered in production agriculture, agricultural mechanics, horticulture, forestry, agricultural products, agricultural supplies and services, and renewable natural resources. The programmes are designed to prepare students to enter as well as to advance in agricultural occupations.

Such programmes had their beginnings before 1900, when a few elementary and secondary schools taught general agriculture as part of the curriculum. During the first decade of this century, serious discussion and experimentation in vocational education created a national awareness of the need for occupational training. Some colleges and normal schools taught agriculture as a general biological science to students preparing to teach general subjects. Practical agriculture subjects with economic utility were added in some secondary schools to induce more boys to complete secondary school and thereby acquire a better general education.

The stated purpose of the Smith-Hughes Vocational Education Act of 1917 was 'to fit for gainful employment'. In the first years, state administrators worked mainly to establish local rural secondary-school programmes. The teachers they hired turned mainly to college textbooks and bulletins for subject-matter content. Supervisors employed by the state government evaluated the classroom, laboratory and field-trip procedures used by teachers and helped them to improve their instruction. The Act

1. Schools in the free public education system, as opposed to private schools.

18

required supervised farm practice by the student. The home project was widely adopted as a method of secondary-school teaching to meet this requirement. The principal goal of early vocational agricultural programmes, therefore, was the preparation of secondary-school graduates for a career in farming.

The influence of scientific farming based on field crop and livestock research at the land-grant colleges made efficient production the aim of instruction in the 1920s. For secondary-school students, the home project was recognized as a first experience in assuming responsibility for livestock or crop production. Teachers were urged to spend time with individual students to help them to learn approved practices in enterprises dominant in the local community. As economic research increased and periods of low farm prices were encountered, emphasis in teaching shifted to management of the total farm business.

Statements of purposes or objectives have been prepared and revised from time to time by committees of the Agricultural Education Division, American Vocational Association, in co-operation with the United States Office of Education. Publications in 1931, 1938 and 1940 were successive editions of Vocational Division Monograph No. 21, *U.S. Office of Education, Educational Objectives in Vocational Agriculture.* They differed mainly in supplementary descriptions. Six major objectives of education for farming involved developing effective ability to (a) make a beginning and advance in farming, (b) produce farm commodities efficiently, (c) market farm products advantageously, (d) conserve soil and other natural resources, (e) manage a farm business effectively, and (f) maintain a favourable environment.

The placement function of occupational guidance was, in the first objective, 'To make a beginning and advance in farming'. Agricultural production, marketing, soil conservation and farm business management were the science-content objectives. 'To maintain a favourable environment' was explained as an effort towards achieving a satisfactory level of family and community living. Social needs during the years of the great depression persuaded leaders to emphasize non-economic purposes.

'Participate in rural leadership activities' became a seventh major objective when Vocational Division Monograph No. 21 was revised in 1955. Until then, leadership development had been included in the sixth objective. The change recognized the education values of FFA and the Young Farmer Educational Association (YFEA) as integral parts of the programme. FFA has provided human relations and leadership training experiences for farm boys in vocational agriculture since 1928. Today it enrols many girls, and the preparation in rural and urban communities includes training for all types of agriculture/agribusiness employment. Local YFEA are distributed widely in many states. Activity programmes are also provided for wives of the young adult farmer members. Competence attained in group participation gives personal satisfaction. It inspires confidence to make the decisions required in independent farm family businesses. Students acquire skills needed to serve as management team members in agribusiness organizations.

Because Federal interpretations of the Smith-Hughes (1917) and George-Barden (1946) Vocational Education Acts limited agricultural education to preparation for farming, the objectives did not mention agribusiness occupations. Frequently, indirect references were made to the need to guide some production agriculture course students into agribusiness occupations. The subject matter they had studied and their supervised work experience were presumed to have prepared them for agribusiness employment even though it was directed towards farming.

The Vocational Education Act of 1963 stated that 'any amounts allotted (or apportioned) . . . for agriculture may be used for vocational education in any occupation involving knowledge and skills in agricultural subjects'. Professional leaders were ready for a change of this magnitude. They had shared in the writing of the *Report of the Panel of Consultants* used by Congress in formulating the 1963 legislation. An American Vocational Association committee was appointed to draft a new set of objectives. The committee sought and obtained consensus among citizens, industry leaders, administrators, teachers and teacher educators for the changes suggested. In 1965 the United States Office of Education published *Objectives for Vocational and Technical Education in Agriculture* (Bulletin 1966, No. 4). The following is a brief but comprehensive list of the six major programme objectives:

1. To develop agricultural competencies needed by individuals engaged in or preparing to engage in production agriculture.
2. To develop agricultural competencies needed by individuals engaged in (off-farm) agricultural occupations other than production agriculture.
3. To develop an understanding of and appreciation for career opportunities in agriculture and the preparation needed to enter and progress in agricultural occupations.
4. To develop the ability to secure satisfactory placement and to advance in an agricultural occupation through a programme of continuing education.
5. To develop those abilities in human relations that are essential in agricultural occupations.
6. To develop the abilities needed to exercise and follow effective leadership in fulfilling occupational, social and civic responsibilities.

An interpretation of each of these objectives follows.

Objective 1 refers to education for farming and ranching. 'Production agriculture' is the term that farm-organization leaders have asked the United States Office of Education to use in reporting and describing secondary-school programmes for farming occupations. The word 'competency' is intended to imply that a learner can demonstrate the correct use of the knowledge and skills learned when provided with similar conditions.

Objective 2 supports education for employment in agribusiness industries and renewable natural resources businesses and services not classified as production agriculture. Teachers of agribusiness courses obtained their preparation for teaching in colleges of agriculture and through actual employment experience.

Objective 3 is occupational guidance to assist students to develop awareness of the agricultural industry, to explore career opportunities, to make plans for and carry out programmes of educational preparation, and later to make needed career adjustments. Guidance involves decision-making about courses to take. It means assisting students to obtain occupational information and exploratory experiences. Supervised part-time employment for wages, called co-operative education, may provide the student with valuable experiences through frequent consultation with the employer and the instructor. Teachers must have special skills in guidance and help secondary-school counsellors to become more effective in providing services to students in vocational agriculture.

Objective 4, occupational placement, is a function that secondary-schools, private technical colleges, the public two-year community colleges, and universities provide for graduates. Private post-secondary business, trade and technical schools attract students largely by guaranteeing jobs for those who complete the training. State Employment Service (SES) local offices annually place seasonal or part-time workers in agriculture/agribusiness jobs. These may include secondary-school and college students. The SES offices are in charge of selection and placement of those who leave school before completion of a degree and under-employed adults who may be enrolled in government-subsidized training programmes. Persons who enrol for continuing adult education courses in agriculture are likely to profit from the added learning and from the guidance and placement services.

Objective 5 assumes that developing the personality to its greatest potential requires certain essential skills in human relationships of all those in an agricultural occupation. Modern business structure demands that each worker be an effective member of a team including managers, supervisors and other employees. The involvement offer extends beyond the circle of economic-oriented associates to interaction with customers, clients and volunteer services in the community.

Objective 6 is related to the development of leadership and the art of following. Individuals who are skilled in human relationships are likely to be able to function well in co-operative roles. Greater depth of insight and breadth of experience in leadership are attainable outcomes of participation in the varied activities of FFA, YFEA, 4-H and other school and community organizations while enrolled in an education programme in agriculture.

Such a major change in the vocational agriculture programme was possible because of the high degree of local teacher involvement at all stages. The teacher's input increased the acceptance of such a major change. Teachers are asked to make use of local advisory committees in planning and conducting a local vocational agriculture programme. In the same way, state directors and teacher education departments use vocational agriculture teachers to help develop objectives and establish standards of performance.

The results of introducing the new objectives for vocational agriculture can be seen from an inspection of student interest as shown in enrolment figures for vocational agricultural programmes. Table 3 gives no comparison with pre-1966 years, but it shows the present diversified enrolment.

TABLE 3. Vocational education enrolment in agriculture by level and type of instruction in the United States

Levels and types of instruction programmes	Enrolment by years		
	1974	1975	1976
Secondary-school programmes			
Production agriculture	328 713	342 556	339 192
Agribusiness			
Supplies and services	22 919	22 117	21 506
Mechanics	104 474	105 149	114 276
Products processing and marketing	7 566	7 880	12 296
Horticulture	55 865	66 947	75 328
Renewable natural resources	15 231	18 021	19 270
Forestry	16 299	15 311	15 991
Agriculture/agribusiness, other	107 231	92 549	114 325
Post-secondary programmes	47 458	59 125	67 663
Adult-education programmes	269 281	282 940	279 870

Source: USOE National Center for Educational Statistics, 1978.

Instructional areas in vocational agriculture

Major instructional areas in vocational and technical education in agriculture are organized on the basis of groups of occupations requiring competence in specialized agricultural science fields. They are: (a) production agriculture (farming and ranching); (b) agricultural supplies/services; (c) agricultural mechanics (sales and services); (d) agricultural products, processing and marketing; (e) horticulture; (f) forestry; (g) renewable natural resources; and (h) agriculture/agribusiness. Each area may be a separate curriculum.

Production agriculture

From 1917 to 1964 the term 'vocational agriculture' was used to describe rural secondary-school programmes for establishment and advancement in farming. There were no other major instructional areas or curricula. The course content varied with the crop and livestock enterprises of communities, regions or states. Classes were organized for secondary-school students in grades nine through twelve. There were continuing-education classes for out-of-secondary-school young farmers and for older adult farmers. Now known as production agriculture, this type of class continues to serve the largest total number of students enrolled in vocational agriculture.

Production agriculture may be defined as an organization of subject matter and learning activities concerned with principles and practices in the production of livestock, field crops, fruits and vegetables, fibre, and other crops on commercial and part-time farms. The curriculum provides general instruction in animal science, plant science, farm mechanics and farm business management, in addition to specific instruction for the students' production enterprises. Knowledge and skills taught involve the economic use of agricultural land, labour, capital and management. The efficient operation of modern farm equipment and the harvesting and marketing of high-quality products are important factors in which students develop skills and technical knowledge. Examples of occupations in agricultural production are: general farmer, livestock farmer or rancher, dairy farmer, fruit grower, farm manager and farm-equipment operator.

The dairy cattle, beef cattle, swine, sheep and poultry enterprises employ a very large proportion of agricultural production workers, predominantly self-employed owner-operators who use knowledge and skills in animal science. Animal-science principles and practices in nutrition, genetics, physiology, animal health, production management, marketing and related areas are instructional units in the curriculum.

The production of field and forage crops, tree fruit and nut crops, small fruit crops, vegetable crops, farm forestry products and other crops together employ the agricultural production workers who use knowledge and skills in plant science. Important plant science instructional units in an agricultural production curriculum are: soils, plant nutrition, plant genetics, plant physiology, pest control, production management, and marketing and related areas. Most producers of meat animals and dairy products grow grain and forage crops for feeding beef cattle, swine, sheep and dairy cattle on the farm; they therefore need technical knowledge and skills in plant science.

For several reasons farm mechanics is treated as a separate element in the production-agriculture curriculum. It deals with physical science knowledge and skills. Modern farms must have a large investment in power and equipment, structures and automation devices to achieve competitive labour efficiency. Instruction in the selection, safe operation, maintenance and repair of machinery and equipment contributes substantially to lowered costs of production. The schools offer instruction in farm mechanics, using well-equipped laboratories called the agricultural mechanics shops. It is feasible to teach this phase of agricultural production at school using tractors, farm machinery and other equipment from the home farms of students, or loaned by dealers, to provide learning experiences that duplicate the abilities needed from day to day in a farm-service centre. These relate to farm power and machinery, farm structures and conveniences, farm electrification, the mechanics of soil and water management, maintenance mechanics, welding, concrete construction, uses of electricity, materials handling, systems development, and other applications of mechanics in agriculture.

Farm business management is basically decision-making. It involves the manipulation of production inputs to achieve chosen goals in terms of output volume, quality and efficiency. The units of instruction are farm accounts, performance records, budgeting and analysis, purchasing, marketing, financial and legal management, farm organizations, government programmes and other applications of economics and business practices. This phase of agricultural education is most appropriate for adult farm owner-operators and farm managers. Careful selection of course content will introduce secondary-school students to management principles and operations.

To supplement the classroom and laboratory learning experiences, each student is expected to develop an agricultural project related to his or her occupational interest. This project could begin with some feeder pigs during the first year and over the four years of secondary education develop into a major swine operation. Another example would be to begin with a few acres of maize on ground rented from the student's parents and expand into a major family partnership over the four years. Such experiences allow the students to practise the knowledge and skills being taught by the vocational-agriculture teacher. During the time the student is planning and conducting the project, the teacher is making periodic visits to provide assistance. This provides an opportunity for the teacher to become familiar with the areas which need greater emphasis in the curriculum.

Work experience programmes are also planned and conducted for students with occupational goals in agribusiness. Arrangements are made for the student to work in an agribusiness to learn specific skills which cannot be taught in the school facilities. These skills are carefully identified by the employer, the student and the teacher. The safety and ability of the student are important factors to consider in developing such a training programme.

Agricultural supplies and services
Businesses that furnish production needs and services to farmers deal in specializations and combinations of manufacturing, sales and services. The principal physical supplies purchased by farmers are agricultural chemicals, livestock feeds, farm-crop seeds, crop fertilizers, petroleum and other supplies, including small equipment. Usually the business that handles supplies for farmers will also furnish services such as grinding, mixing, conditioning and application. Examples of occupational titles in agricultural supplies in which workers need knowledge and skills taught by schools in courses in agriculture are: (a) agricultural supplies manager, (b) agricultural chemicals fieldman, (c) seed salesman, (d) fertilizer applicator, (e) agricultural supplies installation and service mechanic, and (f) feed-mill equipment operator.

An agricultural supplies business may be a local, individually or family-owned enterprise. Today, many are units of state-wide, regional or national corporations. Course content for students employed in or preparing to enter the agricultural supplies field should combine agricultural education with business education. Managers, fieldmen and salesmen who deal directly with farmers, particularly those who give consulting service, need considerable knowledge of animal science, plant science and agricultural business management.

Competency ratings by employers interviewed in state surveys show a high need for general business skills, salesmanship and employee relations by managers and salesmen, and a medium need by service workers. Courses of study in agricultural supplies, sometimes labelled agribusiness, have been developed to the highest level of specialization in non-degree programmes in community colleges and area-vocational technical schools.

Agricultural mechanics

This area of instruction is important enough to be a curriculum specialization in area-vocational schools, technical institutes and community colleges whose graduates are needed in farm-machinery dealerships in regions of high farm production. It covers sales and service of agricultural power units, mostly tractors, integrated machinery and related equipment. Examples of job titles are: (a) agricultural mechanics service manager, (b) agricultural machinery salesman, (c) agricultural mechanics partsman, and (d) agricultural machinery mechanic.

Some positions in agriculture and agribusiness require knowledge and skill in mechanics. Therefore, as a curriculum area, agricultural mechanics sales and service is organized to provide short unit courses to students whose occupational objectives are, for example, horticulture, agricultural supplies or renewable natural resources.

Agricultural products

After farmers have produced quality products, the modern food industry is organized to perform many services and operations in assembling, sorting, testing, grading, processing, manufacturing, storing and marketing. Some of the functions maintain the quality of the product; other operations add value. The numbers of employees in off-farm agricultural products processing and marketing businesses who need knowledge and skills customarily taught in food technology courses in colleges and technical schools vary with the product and with the responsibilities assigned to job titles in the business. The numbers are large at the points closest to direct dealing with producers. After finished products reach the supermarket, or other retail trade outlet, relatively few employees need agricultural education for maximum service to their organizations.

The major food product areas in an agricultural products processing and marketing curriculum are: (a) meat, poultry and eggs; (b) dairy products; (c) fruits and vegetables; and (d) grains for food. Examples of food marketing occupations in which technician-level knowledge and skills in agriculture are used are meat processing manager, fruit and vegetable market manager, livestock buyer, dairy processing equipment operator, grain elevator operator, agricultural commodity grader and quality control technician. Major non-food agricultural products are cotton, tobacco and wool.

Courses in the curriculum include the physical and chemical properties of foods, composition and ingredients of processed foods, formulations and additives. The methods of preservation as related to quality and consumer preference influence the actual practices of canning, pasteurizing, freezing and other operations. Quality control instruction includes not only proficiency in making tests but also requires knowledge of regulations and should give the learner an understanding of the microbiological and other changes that must be controlled. For some employees business abilities are very important; for others special training in mechanics is of value. Adult education for present employees, by increasing their worth to the business, may also aid in raising the job entry qualifications and wages.

Horticulture

Three types of businesses and services that produce, distribute, and utilize horticultural plants for ornamental values are: floriculture, nursery management, and landscaping and turf establishment and management. Greenhouse production and sales, nursery production and sales, garden centre sales and services, landscaping, grounds-keeping, greens-keeping and arboriculture are occupational areas that relate to the types of businesses. Some courses in horticulture will serve several or all of the occupational areas, but others have to be specific to the type of product or service. Examples of horticulture job titles are florist, greenhouse manager, nursery grower, garden centre salesman, landscape aide, greensworker and tree pruner.

School greenhouses are used to teach skills in flower growing and nursery propagation. A headhouse is used for the mixing of soil and potting of plants and workshop facilities are needed for skill development in agricultural mechanics relating to horticulture.

Many persons are hired on a seasonal basis to work in each of these areas of horticulture. Such jobs are desirable for youth, but provide insufficient employment for adults. Owners and managers of nurseries and garden centres are diversifying their businesses to find productive work for year-round employees.

Forestry
The central function of technical education in forestry is to prepare workers for the management of trees grown as a crop. Other aspects of employment at less than the professional level are in forest protection, logging, wood utilization, special products production and co-operation with persons whose work is in conservation or recreation. Some occupational titles in forestry for which technical education is appropriate are forestry aide, Christmas-tree grower, sawmill operator, logger and log scaler.

Renewable agricultural resources
As a curriculum, renewable agricultural resources is an organization of subject matter and learning activities designed to provide opportunities for students to study principles and processes in the conservation and improvement of environmental resources such as forested and other natural areas; fish and wildlife; soil, water and air. It is also concerned with the establishment, management and operation of outdoor recreational facilities. Examples of renewable agricultural resources occupations in which vocational and technical education in agriculture may be used are: (a) recreation farm manager; (b) soil conservation aide; (c) wildlife conservation officer; (d) fish hatchery worker; (e) game farm worker; and (f) park worker.

The course content and supervised practice activities appropriate to preparation for entry or advancement in occupations in conservation, recreational utilization and services connected with agriculturally-related natural resources are structured to treat the multiple uses of forested and other natural areas, and wildlife management, including fish farms and hatcheries. Conservation and utilization of soil, water, air and other resources are taught in the context of regional planning for public, industrial and home owner benefits. A knowledge of legal aspects is important. Most employment opportunities are in government and have regulatory functions to be performed. Many students majoring in other agricultural occupations will profit by at least one course in resource management.

Agriculture/agribusiness, other
No formal curriculum has been prepared or proposed for secondary-school students whose occupational objective is to prepare for a professional position in agricultural fields in industry, government, education or other services. If one of the previously outlined curricula is closely related to the higher-education goal of the student, he/she should enrol for selected courses in it while scheduling adequate college preparatory required courses. The same advice should be given the student who plans to complete a two-year associate degree or technical-school programme in a field of agricultural technology. This is consistent with the introductory discussion in this chapter suggesting that adequate guidance will help each student to have an individualized curriculum. The school schedule should provide the needed English, mathematics, science and social

studies for the vocational agriculture student. Agriculture today requires individuals who are interested and knowledgeable in new agricultural technology and who can keep their knowledge up to date.

The community-oriented learning experiences of a well-taught course in agriculture broaden the acquaintance of the student with organizational leadership patterns and give him the practical life experience needed. Most small schools will be able to offer only two or three agricultural occupation majors. Students whose occupational goal is not in an area offered by the school may elect courses and combinations of courses closely related to the preparation needed. With this background plus work experience in the specific occupation, the preparation should be sufficient for entry level employment.

Future Farmers of America

FFA is the national organization of, by and for students and former students of vocational agriculture in public schools qualifying for Federal reimbursement under the National Vocational Education Acts. The states, territories and other subdivisions of the United States have associations chartered by the national organization. The basic unit in a school is the local chapter of FFA. Local chapters are chartered by and affiliated with an association through which the purposes, rules and programme of activities of the national organization are adopted and supported.

The objects and purposes of the FFA, as revised, include the following:
1. To develop competent and aggressive agricultural leadership.
2. To create and nurture a love of agricultural life.
3. To strengthen the confidence of students of vocational agriculture in themselves and their work.
4. To create more interest in the intelligent choice of agricultural occupations.
5. To encourage members in the development of individual occupational experience programmes in agriculture and establishment in agricultural careers.
6. To encourage members to improve the home and its surroundings.
7. To participate in worthy undertakings for the improvement of the industry of agriculture.
8. To develop character, train for useful citizenship and foster patriotism.
9. To participate in co-operative effort.
10. To encourage and practise thrift.
11. To encourage improvement in scholarship.
12. To provide and encourage the development of organized recreational activities.

FFA's primary aim is the development of agricultural leadership, co-operation and citizenship. The local FFA chapter in a secondary school is an organization of students enrolled in vocational education courses in agriculture. An annual programme of activities is planned and carried out by the students themselves with adult counsel and guidance. The school provides teacher time, facilities, funds and other essential support for FFA as an intra-curricular part of the total personal development of a group of its students.

FFA appeals to young people moving through adolescent years towards adulthood. The opportunity to identify with others whose occupational goals are in the broad areas of the agricultural industry is exemplified in the national blue jacket worn by members. To win medals, trophies, prizes and other awards may seem to adults to have only extrinsic value but to young men in FFA an honour earned through striving for excellence in a worthy activity is at the time highly meaningful. Such an achievement often stimulates the setting of new goals and motivates further study and practice. Public-speaking contests and demonstrations of parliamentary procedure in the conduct of a business meeting have developed in many students a practical, serviceable level of communications competence. In 1978 more than 600 major business organizations throughout the country contributed approximately $900,000 to the National FFA Foundation to be used for an extensive system of awards—national, state and local—that recognize achievements of individual students in agricultural production and in related endeavours.

Group-dynamics techniques need to be understood by teachers who serve as advisers to local FFA chapters. Each member must learn to perform in a variety of roles. The formal ritual of FFA degree ceremonies provides a framework for systematic advancement from year to year. To serve on a committee, to be the chairman of a committee and to be elected as an officer represent stages in leadership development. Planned programmes of officer training for specific duties have been found to be effective in teaching and are often scheduled on a district basis involving several schools at one time. State and national student officers are frequently invited to serve as resource persons. Community leaders in business and professions also assist in leadership training sessions and have a lasting influence on the students. They recognize that the student organization is providing experience that will be of great value to each youth in later years when he joins adult agricultural organizations and other community service groups.

Capable leaders and interested members must have a carefully planned annual programme of work and a calendar of activities. Major subdivisions of the written programme of work of a local FFA chapter generally include the following headings: supervised farming or other occupational experience, co-operation, community service, leadership, earnings and savings, conduct of meetings, scholarship, recreation, public relations and participation in state and national activities. In summarizing detailed suggestions for constructing the programme of work, Bender, Taylor, Hansen, and Newscomb (1979, p. 81) have offered this advice:

Probably the best way to involve all of the members in planning the program is to organize committees to work out areas of the program and present these to the total chapter membership for approval. Some chapters have tried other methods, such as letting the Executive Committee and advisor prepare the program. Some chapters have tried to have all the members attend a leadership training camp in the summer where part of the agenda included working out the next year's program. For the most part, no plan has proved satisfactory unless all of the members participate in its development.

The committees that planned the activities should be given the responsibility for supervising their execution and for the evaluation of the results. Counsel by the teachers of agriculture will assure the contribution of each project or event to the educational objectives of the course and of the school.

Primary level of agricultural education

Agricultural education at the primary level, focusing on an awareness of our food sources, varies from formal classes taught by vocational agriculture teachers to those taught by elementary-school teachers. At the primary level agriculture is not a required part of the curriculum, and there is no special Federal funding available to support such a programme. Since it is an optional curriculum, there are no guidelines or regulations identifying its design. As a result, there is a great variation across the United States in such programmes.

An example of an extensive primary-level agricultural education programme is offered at New Holland, Pennsylvania. The programme is offered to sixth-grade students attending the public schools in the district. It was developed to meet the needs of rural students who leave school early to go into farming. The courses included three main topics: safety around the home and farm; conservation of soil, water and natural resources; and sanitation. These topics were included to provide the student with an understanding of their importance and also to develop an interest in learning more about agriculture. The student was encouraged to start livestock or crop projects and to take care of them until marketed. To accomplish this, they were expected to keep records and a regular schedule of visits was maintained by the teacher during this period. This provided opportunities for individualized instruction concerning the project and an opportunity to meet the parents and develop good relations for the programme and the school.

A less intensive programme that is more widely used is one developed by the elementary-school teacher. The objective is usually to provide an awareness of where the food that we eat comes from. Since much of the United States population now lives in urban areas, many students do not understand where the supermarkets get their supplies. Field trips to visit farms in the area and projects such as hatching chickens in the classroom are used to provide an awareness of how and where our food is produced. Many secondary-school vocational agriculture programmes provide experience for primary-level students by bringing animal exhibits into the primary-level classrooms and explaining how they are grown; or by inviting the primary-level students to visit the vocational agriculture school farm to see how the animals and crops are raised.

Continuing education of adults

A complete programme of education in a community is determined by the needs of all of its people at a period in time, and it should be flexible in providing for the dynamic nature of the many interacting social processes. It is important to the future of the community that a complete programme of education be recognized as that combination of desirable formal and informal educative experiences which will best promote the welfare of the individual in his family and societal functions continuously throughout life. It is believed that formal general education ought to be provided to all individuals as long as they continue their interest.

The controlling objective in long-term planning of young adult vocational education programmes in agriculture is the continuous individual progress made towards becoming established in an agricultural occupation and in successful family and community living. The secondary school provides an introduction to the knowledge of skills needed for such a goal. In most communities in which the teacher of agriculture is doing outstanding work in advising and instructing young adult farmers and other persons employed in off-farm agricultural occupations, there is also a superior secondary-school student programme in agriculture. Frequent involvement in the current problems and needs of employed young adults provides teachers with the best preparation for organizing appropriate learning activities for secondary students. The example of systematic instructional contacts by their instructors with adults in agricultural occupations over a period of years after completion of secondary school encourages younger students to approach occupational development planning with vision, confidence and persistence.

Prior to 1963, leaders in agricultural education designed the training pattern for adult classes in agriculture by combining on-the-job instruction with attendance at regular classes. The instructor taught the class at least four hours per week and used the other part of the day to visit the adult students on the farms they operated or where they were employed, and provided a planned sequence of on-the-job learning experiences. The demonstrated efficiency of this combination pattern of instruction later encouraged school districts to assign portions of daytime working hours of the teachers of agriculture to be used in individual on-farm instruction. Since the enactment of the Vocational Education Act of 1963, the procedure is being applied to adult education for off-farm occupations.

Increased capital requirements in agricultural production, reduction in numbers of farm workers and more complex involvements in marketing have focused attention on management as the prime educational need of persons who own and operate modern commercial farms and allied agricultural enterprises. Technical information relating to seeds, feeds, fertilizers and agricultural chemicals is readily available from manufacturers and distributors of these supplies. Dealers in farm power and equipment have learned that it is very worth while for them to assist in instructing their farmer customers in the selection, safe operation and preventive maintenance of each tractor, implement or other item in agricultural mechanization and automation. Vertical integration in the marketing and processing of food and other agricultural products has brought another large segment of American business directly into educational relationships with farmers. Income tax regulations have made the keeping of complete farm accounts a necessity. These factors have resulted in management education becoming an important part of the curriculum in adult classes.

In an Office of Education bulletin prepared in collaboration with vocational educators in states with many young adult farmers enrolled in post-secondary-school classes in agriculture, Hunsicker (1956, pp. 4–6) presented the imperative requirements for continuing education in farm business organization and management in this way:

When young farmers leave or graduate from secondary high school they soon find that their needs and problems have multiplied. Those who are considering farming as an occupation will have to analyze and re-examine their interests, intentions to farm, and opportunities to become established as farmers. . . . Even those fortunate enough to start with a farm, a minimum of machinery and equipment, and a will to succeed face difficult problems and choices.

As young farmers progress toward successful establishment in farming, they will recognize the need for instruction in: Developing parent-son agreements in farming, renting farm land, locating available finances, producing farm products efficiently, selecting and maintaining farm equipment, marketing farm products, keeping and analyzing records, developing farm and home plans, planning land use and conservation programs, laying out crop rotation systems, interpreting government programs, interpreting and executing legal papers, making tax returns and Social Security payments, and participating in farm and community organizations. Further education and training will develop the ability of young farmers to better solve many of their perplexing problems in these areas.

To implement the foregoing concepts, a number of states gave official endorsement to 'the farm management approach' to adult education in agriculture. The support was in various forms: additional agriculture instructors were hired, time was assigned to adult classes including individual instruction of students and additional state funds were allocated to schools that conducted approved programmes.

The situation at present is such that every available technique for further improvement in management education and in technological training of commercial farm operators needs to be thoroughly applied in existing programmes. An added supply of qualified teachers is needed to staff adult programmes in more secondary schools. Specialized programmes will be initiated in area vocational-technical schools as they are built with funds provided under the Vocational Education Act of 1976. Since most of the present vocational education programmes in agriculture are located in local school districts with concentrations of commercial farms, it is likely that adult instruction in agricultural production will continue to expand in these schools.

Administrative relationships

Having an understanding of and giving approval to a basic philosophy—adults can learn and want to continue to learn—is fundamental. Having an awareness that adequate physical facilities, in terms of public school buildings and equipment, are available at little extra cost is a practical consideration. Finally, the chief school administrator and local board of education must be able and willing to accept leadership responsibility in organizing and administering appropriate adult programmes.

In the past in rural secondary schools the initiative for organizing adult classes for farmers has come from the local teachers of agriculture. They may have acted from a desire to provide the community with the quality and variety of service that had been described to them as a model in their professional education in the state universities, or they may have recognized that former secondary-school students urgently needed continuing instruction.

Most of the details of administration may be handled according to established policies in the school. To qualify for state financial subsidies approval must be obtained from vocational-education supervisors. Explanation of local need for flexibility, interpretation of standards, or requirements normally results in approval being given and may cause the innovative features to be treated as a pilot programme. At a future date, the dissemination of successful procedures to other schools may inspire further advance in the local programme.

Adult vocational education in agriculture in public schools in based on voluntary enrolment. Fees charged are low. Regular attendance and completion of courses depends upon student appraisal of the value of the instruction. Surveying the needs of potential students and getting to know them personally are, therefore, essentially as much a part of administration as of instruction.

Class teaching

The reasons why people attend voluntary, non-college credit adult classes are important to administrators, to adult education programme co-ordinators and to instructors who must select the learning activities and use the most effective teaching procedures. Not least, they ought to be understood by the students themselves. Jensen and others (Jensen et al. 1964) have explained that the first interest of youth education in the schools is in socializing the child, and that the adult educator faces the task of re-socializing the adult. They observed that in a static traditional society adult education is redundant; there is no need to change social roles or acquire new skills. In our society, technological change demands new skills and shapes human relationships; social mobility permits and encourages changes in status, in values and in social relationships; geographical mobility generates necessity for adaptations to different modes of living and to new economic situations. The worlds of the adult, in general order or priority, are the world of work, the social world, the world of form and the world of nature. Vocational education skills can be clearly specified and thus efficiently learned. Social competencies also need real settings in which to be developed.

Mason and Haines have stated that for out-of-school youth and adults increased job competency in their present occupations or preparation for different employment comes to some degree from on-the-job training, but increasingly it comes from organized courses offered by schools and by employers and their trade and professional associations (Mason and Haines, 1965, pp. 300–302). A basic approach to programme planning in adult education is the analysis of people and their occupational needs. Individuals, at one time or another in their lifetime, may need continuing adult education in preparation for a new occupation, to increase performance in the present job, in preparation for advancement, or as retraining made necessary by a variety of factors. Needs may be classified by job level into employee, supervisory and management. Mason and Haines (p. 302) continued their discussion of the 'needs approach' with the following:

The coordinator engaged in adult education should be aware that some adults enroll in vocational classes to meet needs not expressed in the occupational classification already discussed. A great many people enroll to satisfy needs such as the desire to associate with other people in the same occupations; the need for stimulation to their thinking; the wish to increase status with fellow employees, supervisors, friends, and family, which comes with increasing their educational level; or the desire to gain certificates attesting to their accomplishments.

There are certain essentials in a teaching plan. The total unit of instruction should be clearly specified and its component problem areas delineated. The number of class sessions to be devoted to each area or topic must be determined, subject to modification during the course as progress of the group is found to be faster or slower than anticipated. The session topics or lessons should be used as an integral part of the learning process. With appropriate consideration for the much wider experience of mature students, the principles of learning for use with secondary-school classes apply to adult groups. Group discussion techniques are admirably suited to management problems. Demonstrations and field trips reduce misunderstanding of unfamiliar terms and confusion caused by imprecise verbalization.

Young Farmers Association

Leadership development is a very important adult education objective. Participating experiences are required to achieve it. Each individual must have real responsibilities in situations that involve personal and group goals that have value and meaning. Co-curricular activities in secondary school are desirable but limited to the type of social control in which teachers and parents occupy an authoritative role. A local school Young Farmers Association whose members are the persons enrolled in the post-secondary-school classes in agriculture today has important leadership training functions in many communities in an increasing number of states. Indiana, Ohio, Virginia, Pennsylvania, Texas, Utah, California, South Carolina and some others have had more than twenty years of experience with state associations of affiliated local units, sometimes called chapters. Membership is open to everyone enrolled in the adult education programme in agriculture in the school.

The Young Farmers Association operates with a set of by-laws, has a definite officer and committee structure, and prepares and carries out a written annual programme of activities. Prominent in the goals of the organization are items that support the programme of class instruction; the activities relate to ways in which the members may assist the teachers of agriculture with their instructional responsibilities. The Association may raise funds for teaching materials. Committees are assigned to obtain services of resource persons. Arrangements for transportation on field trips and tours are handled by the class members. The executive committee is a very fine department advisory council. In a growing number of instances, one member accepts responsibility for individual instruction of another. This is a planned aid to diffusion of improved practices and new knowledge in the community.

A young adult student organization is related to individual on-farm or on-the-job instruction more in terms of its contribution to improved family and community living than as a requirement in instruction leading to advance in occupational skill and efficiency. A Young Farmers Association programme of work generally includes community service activities, co-operative activities and social and recreational activities. Wives of the members participate in some or most of the projects, meetings and events. Training received frequently results in the young farmer becoming a member and leader in state and national farm organizations and in regional civic groups.

The agriculture classroom

Flexibility of arrangement of student tables, chairs and other equipment is a dominant feature of classrooms for secondary-school and adult classes in agriculture. Ample shelves, cabinets and other types of storage space for reference materials, bound and unbound, occupy interior wall areas or are located in adjacent storage rooms. This accommodates exhibits, demonstrations and role-playing efforts, and allows physical separation for small group or committee work activities. Built-in convenience outlets and other structural details promote efficient use of all modern types of audiovisual media.

The agricultural mechanics shop

The agricultural mechanics shop has served primarily for instruction of students and adults in the selection, safe operation, maintenance and repair of agricultural production machinery and equipment. As additional specialized programmes for off-farm agricultural occupations are organized, adjustments in the teaching of mechanics principles and practices appropriate to supplies merchandising, products marketing, ornamental horticulture services and resources management will be made. The first precaution in designing a shop is to be sure that it is large enough to allow for adaptation to new emphases.

Hollenberg and Johnson (1960, p. 32) assembled detailed suggestions for many construction features that contribute to superior instruction in agricultural mechanics, among which are:

The shop is preferably an integral part of the same structure in which the classroom is located, and is deserving of the same consideration as to esthetics, engineering, and economics in the selection of building materials, architectural design, and equipment. . . . Two features are of paramount importance: (1) large areas of unobstructed floor space, and (2) easy access provided for large farm machines or other projects to and from the shop. Agricultural mechanics shops vary in size, but 40 feet is a minimum width. A width to length ratio not greater than 1 to 2 is desirable.

It is important that the ceiling in a shop be high enough to provide headroom for large farm equipment. . . . Usually a 14-foot ceiling will be adequate. . . . The entry door should be 14 to 16 feet wide, preferably of the overhead type.

Many other features were listed by Hollenberg and Johnson. Several states maintain regularly revised lists of equipment for agricultural mechanics. Stress is placed variously on the economy of fenced-in and partly-roofed outdoor storage areas, an unloading ramp, colour dynamics, safety measures including special welding, spray painting and steam-cleaning areas.

The conference-office-laboratory area

In the modern concept of adequate housing for a complete programme of vocational-technical education in agriculture, an area nearly equal in square feet of floor space to that of the classroom is needed for conference, office and laboratory functions. Adult students make frequent use of such a combination room or area during daytime hours when secondary-school classes are meeting in the regular classroom and shop. When a committee of adults or of secondary-school students decides to work on a class or student leadership activity, they should have a conference table area available to them. Any student may arrange to do individual testing of soil, seed, grain or other crops, milk or other livestock products, or to use desk calculators and other office machines in the laboratory area. Planning activities for work to be done in the shop may be more conveniently worked on in the conference room located between the classroom and the shop. Other facilities such as a greenhouse or food technology laboratory are required for specialized programmes.

Bibliography

BARTON, M.; HOLMES, G. E.; BUNDY, C. E. 1963. *Methods in Adult Education.* Danville, Ill., The Interstate Printers and Publishers, Inc.

BENDER, R. E.; CUNNINGHAM, C. J.; McCORMICK, R. W.; WOLF, W. H.; WOODIN, R. J. 1972. *Adult Education in Agriculture.* Colombus, Ohio, Charles E. Merrill Publishing Co.

——; TAYLOR, R. E.; HANSEN, C. K.; NEWSCOMB, L. H. 1979. *The FFA and You.* Danville, Ill., The Interstate Printers and Publishers, Inc.

Educational Objectives in Vocational Agriculture. 1940 [rev. ed. 1955]. Washington, D. C., Office of Education. U.S. Government Printing Office. (Vocational Division Monograph, 21.)

HOLLENBERG, A. H.; JOHNSON, E. J. 1960. *Buildings, Equipment, and Facilities for Vocational Agriculture Education.* Washington, D. C., Office of Education, U.S. Government Printing Office (OE-81003).

HUNSICKER, H. N. 1956. *Planning and Conducting a Program of Instruction in Vocational Agriculture for Young Farmers.* Washington, D.C., U.S. Department of Health, Education, and Welfare, Office of Education. (Vocational Division, Bulletin No. 262.)

JENSEN, G. A.; LIVERIGHT, A. A.; HALLENBECK, W. (eds.) 1964. *Adult Education: Outlines of an Emerging Field of University Study.* Chicago, Ill., Adult Education Association of America.

MASON, R. E.; HAINES, P. G. 1965. *Cooperative Occupational Education and Work Experience in the Curriculum.* Danville, Ill., The Interstate Printers and Publishers, Inc.

National Center for Educational Statistics. 1978. Washington, D.C., Office of Education, U.S. Government Printing Office.

Objectives for Vocational and Technical Education in Agriculture. 1966. Washington, D.C., Office of Education, U.S. Government Printing Office. (Bulletin No. 4, 1966.)

Official Manual Future Farmers of America. Alexandria, Va., Future Farmers Supply Service. (Revised periodically.)

Summary of Research Findings in Off-Farm Agricultural Occupations. 1965. The Center for Vocational and Technical Education, Columbus, Ohio, the Ohio State University.

Post-secondary agricultural education in colleges and technical schools

Edgar Paul Yoder

Post-secondary education in agriculture includes all instructional programmes in agriculture-agribusiness and renewable natural resources beyond the secondary school level but which are less than a baccalaureate degree (Sidney, 1978). The term post-secondary education may carry multiple meanings, depending in part on its contextual usage. A definition of what constitutes a post-secondary institution was required by the Congress of the United States under terms of the Education Amendments of 1972. Accordingly, a post-secondary institution was defined as:

An academic, vocational, technical, home study, business, professional, or other school, college or university, or other organization or person offering educational credentials or offering instruction or educational services [primarily to persons who have completed or terminated their secondary education or who are beyond the age of compulsory school attendance] for attainment of educational, professional or vocational objectives.

This definition (cited in Riendeau, 1975, p. 153) excludes from consideration as a post-secondary institution: (a) organizations that provide on-the-job training, apprenticeship training, or formal instruction in a business organization for its own employees or the employees of customers; and (b) secondary institutions that offer adult education courses.

Post-secondary agriculture programmes are offered through several types of institutions, including technical institutes, junior colleges, community colleges, area vocational schools, proprietary trade and technical schools, and four-year universities and colleges. Examination of data in Table 4 reveals that most post-secondary agriculture programmes are offered through community college institutions, whereas less than one-half (44 per cent) of the states offer post-secondary agriculture programmes in the land-grant universities. Prior to 1930, many land-grant universities offered short courses and one- and two-year programmes in agriculture. Following the Second World War they decided upon a shift in mission towards baccalaureate-level basic science and graduate study. For several years Veterans On-Farm Training, offered in most counties of all states, replaced the formerly popular two-year and other short courses in production agriculture offered at land-grant universities.

Development and expansion of post-secondary programmes

Beginning in the 1920s and hastened by the Second World War, the agricultural industry encountered a rapidly expanding scientific and technological base. In addition, an increasing proportion of agriculturists recognized the need, within a 'technologically exploding' agricultural sector, for formal agriculture programmes beyond the secondary

TABLE 4. States and institutions with two-year post-secondary education programmes in agriculture and agribusiness in 1977 by types of institutions

Type of institution	Number of states	Number of institutions
Area vocational school	17	61
Technical institute	11	56
Technical college	8	24
Community college	29	148
Junior college	9	30
College	20	119
University	22	40

school level. Some individuals were able to meet their needs by attending short courses, winter institutes and other short duration programmes. The need for additional formal agriculture instruction beyond secondary school, but at less than the baccalaureate degree, was recognized, and courses were initially offered in the New England states and in New York (Sidney, 1978).

The rapid expansion of post-secondary agricultural education programmes was very evident during the 1960s and continued into the 1970s. Table 5 reveals the extent of the phenomenal expansion in the number of institutions offering post-secondary agricultural programmes and the corresponding increase in enrolments.

TABLE 5. Annual summary of institutions offering post-secondary agricultural programes and enrolments

Year	Institutions	Enrolment
1963-64	60	2 800
1968-69	183	11 036
1972-73	401	34 924
1974-75	450	47 458
1976-77	478	67 663
1977-78	513	73 597
1978-79	513	88 291

Sources : The data are derived from Hunsicker, 1979, and Mokma, 1979.

This increase is at least partially attributed to one or more of the following developments:

1. There has been and continues to be a rapidly expanding agricultural, scientific and technological base which necessitates personnel with more specialized capabilities and skills, including mid-management and technical-level competencies.
2. This expansion also includes non-production agriculture occupations as well as production agriculture occupations within the total scope of agriculture/ agribusiness. The base of agricultural occupations is expanding significantly with the inclusion of non-farm agricultural occupations in horticulture, agricultural

mechanics, forestry, renewable natural resources, supplies and services, and product processing and marketing.

3. There is a strong American faith in the value of education and the economic and non-economic benefits that may accrue to those who participate in formal agriculture programmes.

4. Growing out of this faith in formal education programmes has come an employment structure in which many positions require specific credentials. Agricultural businesses, industry and government expect the educational system to 'screen' people. This is done through licences, degrees, certificates or academic achievement (Whinfield and Johnson, 1974). As a result, some jobs involving competencies in agriculture today require formal education beyond secondary school, whereas ten years ago the same agricultural job did not have additional formal education beyond the secondary-school level as a basic requirement.

5. The labour market itself contributes to the increased expansion of post-secondary agriculture programmes. The highest unemployment is in the 16-to-20 age-group; therefore, for some the alternative to unemployment is additional education.

6. The composition of the population is a factor in an increasing demand for post-secondary programmes. Individuals in the segment of the population labelled as the 'post World War II baby boom' in the United States are now approximately 30 years of age. This has contributed to a larger proportion of the population now approaching middle age and making successive career changes that require additional education.

7. The increasing response of the American governmental system to the needs of people has resulted in national and state legislation that supports, in part, post-secondary programmes. The financial support for instruction and supportive services enables post-secondary institutions to be more responsive to the agricultural needs of society.

8. Enrolments in agriculture in state land-grant universities in the 1970s increased far beyond expectation. Awareness that many positions could be filled adequately by persons with two years of post-secondary agricultural education in regionally important enterprise areas contributed to new programmes in agriculture. These programmes were introduced into community colleges, technical institutes and some area vocational technical schools.

9. During the late 1960s and into the 1970s, the American public became more aware of the importance of agriculture in the economy. This is especially evident when the balance of payments of the Unites States is considered.

10. In addition, agricultural profits improved which contributed to a renewed interest in agricultural careers. This in turn heightened the demand for post-secondary agricultural programmes.

Although the rapid technological development in agriculture is undoubtedly a major factor in the expansion of post-secondary agricultural programmes, interaction with the other factors contributes to an increased demand for high-quality programmes in agricultural education at the post-secondary level.

Organization and structure

Co-ordination of post-secondary education programmes is achieved through various types of organizational structures. The overall purpose of the co-ordination function is to regulate and provide leadership necessary for the accomplishment of the goals and objectives of post-secondary education. Gillie (1973) identified three types of co-ordinating bodies for post-secondary education: (a) a single co-ordinating board, which usually is a governing agency; (b) a board that is authorized to co-ordinate and control certain selected activities of the institutions but is restricted in general governance and administrative powers; and (c) a board that is organized strictly as a voluntary system. This latter type of co-ordinating body includes representation from all post-secondary institutions involved in a state and serves to co-ordinate activities that are of common concern to the participating institutions.

Within the United States various organizational structures for control and co-ordination of post-secondary education may be found. In a state, these arrangements may be categorized as follows: (a) a single board for all higher education (including post-secondary programmes of less than a baccalaureate level) institutions in a given state; (b) several boards with each responsible for a specific type of institution (example: a special board for community and junior colleges, a special board for state colleges and a third board for state universities); (c) individual boards of control for each type of institution, with a super co-ordinating board (example: the super co-ordinating board may be designated as a commission for higher education); (d) individual boards for each institution, with co-ordinating units for colleges, universities and community colleges; (e) individual boards for each college and a voluntary co-ordinating umbrella-type commission for the entire state or region within a state; and (f) boards for each individual college and no overall co-ordination between them (Gillie, 1973).

The organizational structure for each specific post-secondary institution within the overall state plan for co-ordination may take varying forms. The specific structure, whether it be a vertical or horizontal organization, and the specific titles of personnel within the structure, are primarily a reflection of the size and programme scope of the institution.

Financing post-secondary programmes

Funds for post-secondary agriculture programmes are secured from several sources —local, state, and federal; students; and private donations (Lombardi, 1973). The total funds for operating post-secondary programmes are determined to a large extent by the size (number of students) of the programme. Funds are typically allocated by the responsible governing body to the post-secondary institution based on the number of full-time student equivalents (FTSE) enrolled in the institution. The amount of funding each respective institution allocates to the agricultural programmes is then, in part, determined by the number of FTSE served by the agricultural programme. However, the number of FTSE for the agriculture programmes is not the sole factor considered in determining the amount of funds allocated to the programmes. Generally, it has been recognized that many post-secondary occupational education programmes, including agricultural education programmes, cost more per unit than do the general and transfer academic programmes. Thus, in making funding allocations for these programmes, the administration will consider other factors, such as agricultural programme cost differentials, in addition to the number of FTSE.

Most locally-controlled post-secondary institutions depend on state subsidies, student tuition fees and local property taxes to provide from 90 to 95 per cent of the funds needed to operate the institution (Lombardi, 1973). The three major local property taxes which may be used to support locally-controlled institutions are: (a) general-purpose property tax typically used for general operations; (b) special purpose tax; and (c) capital-outlay tax for buildings and equipment. With few exceptions, state-controlled post-secondary institutions do not have direct recourse to local property taxes; they rely primarily on student tuition fees and state subsidies for funding.

In some states, specific legislation has been formulated which directly supports post-secondary programmes. In Illinois, the Public Junior College Act provides significant support for technical education programmes (Williams, 1972). The Act created guidelines for the establishment of junior colleges; it stipulated that programmes would include courses in occupational, semi-technical or technical fields leading to employment. Many states currently have legislation that provides appropriations from the state treasury that may be utilized to support post-secondary occupational programmes.

Another major source of funds for post-secondary programmes is from the federal government. A number of legislative acts including the Higher Education Act (PL 89-329), Vo-Tech Student Loan Insurance (PL 89-287), Vocational Educational Amendments of 1968 and the Education Amendments of 1972 are examples of federal legislation which provide funds for post-secondary programmes.

Federal funds for post-secondary programmes are also available through the Department of Labor and the Department of Agriculture. Although the relative proportion of federal government financial support for post-secondary programmes is small compared to state and local funds, the availability of federal funds has been important in the continued development of post-secondary programmes in agriculture.

Curricula and programmes

The rapid change in agricultural technology and the newly identified agricultural occupations have contributed significantly to the large increase in the types of agricultural curricula and specific agricultural courses available through post-secondary institutions.

Types of programmes

Many types of programmes in agricultural education at the post-secondary level are found in the United States. The types of programmes reflect differences in the course requirements, objectives of the programme and length of time required for its completion. The type of post-secondary agricultural programme completed may result in the enrollee receiving an Associate in Applied Science (A.A.S.) Degree, an Associate in Science (A.S.) Degree, an Associate in Arts (A.A.) Degree, or an Associate in Occupational Studies (A.O.S.) Degree. Table 6 presents details that help differentiate the various degrees offered in post-secondary agricultural education programmes.

In addition to the associate-degree types of curricula designed to prepare persons for technical or skilled positions, a variety of other curricula are offered at the post-secondary level. Two of these programmes are the certificate programme and the agricultural short courses. Occupations such as dry kiln operators, pesticide applicators or farriers, among others, do not require an in-depth two-year Associate Degree curriculum. For many of these occupations, all that is required to develop job-entry skills are specialized one-year courses. The certificate type programme parallels the Associate in Occupational Studies Degree to the extent that the content of the curriculum consists of technical agriculture subject matter related to an agricultural occupation. The certificate programme does differ from the Associate of Occupational Studies programme in that the certificate programme takes one year or less to complete. At the completion of the prescribed curriculum, the enrollee is issued with a certificate of proficiency and completion.

TABLE 6. Basic requirements for selected degrees offered in two-year post-secondary agricultural education Programmes, with a total of sixty credit hours required for completion of each degree.

Degree	Credits required in specialized and supporting agricultural courses	Credits required in general studies courses
Associate in Applied Science	60	0
Associate in Occupational Studies	40	20
Associate in Science	30	30
Associate in Arts	12	48

Many land-grant universities are now offering agriculture short courses through a continuing education division. These short courses, usually one to five days in length, are designed to upgrade the worker with regard to new ideas and developments in his occupation. Short courses may also be called workshops, seminars, or conferences. Examples of agriculture short courses and conferences which have been offered include: dairy production short course, milking management seminar, beekeeping short course and lumber grading workshop.

Practical work experience in the curriculum

Many post-secondary agriculture programmes offering the associate degree include as part of the curriculum a practical agricultural work experience requirement. This practical experience requirement may take the form of working on the school farm and/ or completing an occupational internship wherein the student is co-operatively placed in an appropriate agricultural industry. The internship is supervised by an agricultural faculty member, and the student receives a grade for the internship experience.

The school farm provides an opportunity for the faculty to relate the classroom instruction to practical laboratory activities conducted at the school farm. Many school farms have a farm manager who works with the faculty in operating the school farm and co-ordinating class activities with farm work. The school farm manager is advised by a farm operations committee composed of agriculture faculty members. This committee meets frequently throughout the year to plan and revise the enterprise goals and specific activities of the farm.

Articulation of post-secondary curricula

The expansion of agriculture programmes in post-secondary education, secondary vocational programmes and four-year colleges and universities has increased the need for programme articulation between the institutions. The National Advisory Council defined articulation as a planned process which facilitates the transition of students between the secondary and post-secondary levels and permits unhindered student continuity through the various educational levels. Increased efforts must be made to improve the articulation among the agricultural programmes at various levels so that students can move from secondary schools and universities without repeating courses or experiences already mastered. Students are allowed in some instances to earn credit by examination or by advanced placement. An increased emphasis on articulation activities is needed and must be an integral component in programme planning. Future articulation efforts must involve personnel from all levels of agricultural education and personnel from agricultural business and industry.

Post-secondary student organizations

Agriculture curriculum related organizations have been a part of some post-secondary agricultural education programmes for some time. Many faculty members recognize the value of student organizations and their related activities as an integral part of the post-secondary agriculture curriculum. These activities are viewed by most agricultural educators as a 'method of instruction' that enhances the classroom and laboratory learning experience.

Prior to 1979, there was no formal overall national structure for post-secondary agriculture student organizations. In April 1979 a National Conference for the National Post-Secondary Student Organization was attended by representatives of twenty states. The objectives developed for the organization were:

1. To promote recognition of the value of post-secondary vocational technical education.
2. To encourage co-operation among students in the various curriculum areas.
3. To develop leadership and management abilities in agriculture, agribusiness and renewable natural resources.
4. To assist the individual in making his/her occupational choice.
5. To develop character, citizenship and patriotism.
6. To promote a favourable relationship with the business and industrial community.
7. To co-ordinate and co-operate with educational, professional and service organizations.

The faculty in post-secondary programmes

A majority of students enrolled in post-secondary agricultural programmes prepares for mid-management level occupations. The students and the employers of graduates expect the subject matter offered to be of sufficient breadth and depth to qualify the graduate for such positions. The faculty member in post-secondary agriculture programmes must be both a specialist in the technical area he/she is teaching and a competent director of learning.

A unique aspect of the faculty in post-secondary agriculture programmes compared to secondary agricultural programmes is the relatively high proportion of part-time faculty members. Approximately 45 per cent (1,595 out of 3,563) of the faculty in post-secondary agriculture were classified as part-time during 1978/79 (Mokma, 1979). Many of the part-time faculty members are employed in some aspect of the agricultural industry and teach in continuing education agricultural programmes offered by area vocational schools, technical schools and community colleges.

Personnel employed to teach agricultural education in the post-secondary setting have often been secured from one of three sources: (a) teachers currently teaching agriculture at the secondary school level, (b) Master's and Ph.D. graduates in technical agriculture subject matter areas from four-year colleges and universities, and (c) practitioners in the agricultural industry.

One urgent problem is the lack of pre-service teacher education which specifically prepares for post-secondary teaching. Moreover, there is a lack of inservice education programmes to upgrade the pedagogical skills of teachers recruited from industry or four-year colleges and universities. It is essential that the faculty and administration work co-operatively to design a teacher inservice programme to meet teachers' needs, thereby resulting in improved agricultural programmes for students.

Faculty members in post-secondary agricultural programmes are given academic rank based on the individual qualification. Initial faculty appointments are made to the academic rank for which the individual is judged to be qualified as evidenced by previous experience, academic credentials and recommendations. The institution usually identifies minimum standards for each level of academic rank. Typical academic ranks found in post-secondary institutions are instructor, assistant professor, associate professor and, at the highest level, professor. Some institutions use somewhat different titles such as assistant instructor, associate instructor, instructor and, at the highest level, senior instructor. It is not uncommon for post-secondary institutions to offer degree-equivalent credit to the faculty members for their years of professional experience. Professional experience has generally been defined as any combination of trade experience, teaching experience or post-secondary education. Examples of degree-equivalent credit for years of professional experience are: (a) associate degree, two years of professional experience; (b) bachelor's degree, four years of professional experience; (c) master's degree, five years of professional experience.

All faculty members in post-secondary institutions are involved in a systematic programme of evaluation and supervision. Most institutions have a specific system for promotion and tenure that utilizes information from the faculty evaluation and supervision component. Tenure and promotion considerations for faculty members are usually based on: (a) fulfilment of the minimum qualifications for a higher rank, (b) effectiveness in performing assigned duties and responsibilities, and (c) length of service.

Evaluation of post-secondary programmes

Evaluation of post-secondary agricultural programmes assists in the development of programmes that meet the needs of society in general and individuals in particular. Evaluation, in essence, is a procedure in which information is obtained by various means from a source or several sources for making judgements about the course, programme, or institution being examined. The entire process of evaluation is illustrated in Figure 2.

There are two principal approaches to the evaluation of post-secondary programmes, namely, the process approach and the product approach. (Medsker, 1971, p. 9). The process approach in post-secondary agriculture programmes is one that is largely concerned with how the programme functions. The process approach, sometimes referred to as formative evaluation, would assess factors such as the use of advisory or craft committees, how the curriculum is developed, the facilities and equipment available for instructional purposes, the quality of the faculty, the quality of the teaching process and the process for selecting students. The basic purpose of the process approach is to determine how the objectives are accomplished. The product approach attempts to determine how well the objectives of post-secondary agriculture programmes are accomplished. This approach, frequently called summative evaluation, would be based on information regarding the number of students placed in agricultural related occupations, the adequacy with which graduates perform their jobs and the earnings of graduates from a programme.

Information for evaluation purposes may be obtained from a variety of sources,

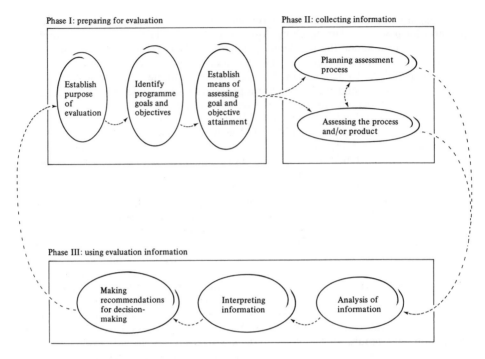

Fig. 2. Evaluation model for post-secondary agriculture programmes.

including programme graduates and employers of graduates. Additionally, evaluations may be completed from within the institution (self-evaluation) or by an external evaluation team or agency. Programme evaluations in post-secondary institutions offering agricultural education programmes involve a combination of self-evaluation and external evaluations. Many states require that programmes be formally evaluated from within every three years, whereas external evaluations must be conducted once every five to seven years.

In recent years, follow up studies have been increasingly used to provide information necessary for assessing the extent to which post-secondary agriculture programmes are reaching the objectives. These studies generally involve asking former students or employers of former students a series of questions designed to provide answers to criterion questions. The following objectives have been identified for conducting former student follow up studies:

1. To determine career patterns of former students.
2. To identify the demand for positions within a region or community.
3. To determine the occupational mobility of former students.
4. To determine the adequacy of the programme in preparing individuals for job entry and advancement.
5. To determine the adequacy of ancillary services. Once the objectives have been identified, specific criterion questions can be formulated to obtain the information necessary to determine the extent to which the objective was accomplished. The objectives and criterion questions provide the basis for determining what group to

follow up, what techniques should be used and the time-frame required to conduct the follow up.

Summary

The phenomenal growth of post-secondary agricultural education programmes in the United States can best be explained through an increased demand for agricultural programmes that are responsive to community and regional agricultural needs. Factors that provide a basis for quality post-secondary agricultural education programmes are (*Agricultural Education for the Seventies and Beyond*, 1971, pp. 23–4):

1. Post-secondary school institutions must seek instructors with appropriate teaching and technical abilities. Dynamic teacher education programmes are needed to supply qualified instructors having expertise in the technical field.
2. The curriculum, in order to be relevant, must be based on the needs of (a) the industries employing the graduates, (b) the students who seek the training, and (c) the society or environment in which graduates are to operate. Continuing contact must be maintained with the agricultural industry through advisory committees and co-ordinated work experience programmes. Systematic evaluation of the success of graduates will keep the curriculum from becoming obsolete.
3. Vocational and technical education should not be considered terminal. Students and educators should understand that continuing education will be normal for agricultural workers at all levels.
4. Post-secondary school programmes should be articulated with secondary-school programmes. Co-ordinated planning with secondary-school teachers and administrators will assist in the development of post-secondary school programmes that build upon knowledge and skills the student has gained in secondary school.
5. Post-secondary school programmes in agriculture should be available to serve the needs of youth and adults at all levels in both rural and metropolitan areas. Programmes must provide educational opportunities for the secondary school dropout, for employees who want to be retrained or upgraded, and for the disadvantaged. They should also serve as a stimulus to the student who may later choose to enter a four-year programme.
6. Occupational experience designed and supervised co-operatively by educator and employer should be an integral part of these programmes.
7. Adequate facilities and modern equipment must be available to provide training for industry needs.
8. Recruitment, placement and follow up of students must be recognized as major components of effective programmes.
9. Programmes should be continually evaluated.

Bibliography

Agricultural Education for the Seventies and Beyond. 1971. Washington, D.C., American Vocational Association.

BENDER, R. E. 1977. The Programme of Agricultural Education with Implications for Colleges of Agriculture and Natural Resources. In: D. L. Armstrong (ed.), *Impact of Enrollments and Student Body Composition on Academic Programs, Design and Delivery.* East Lansing, Mich., Michigan State University.

BYLER, B. L.; WILLIAMS, D. L. 1977. *Identification of Activities to Enhance Articulation Between*

Secondary and Post-Secondary Vocational Agriculture Programs in Iowa. Ames, Iowa, Iowa State University (ERIC Document Reproduction No. ED 147 562).

ERPELDING, L. H. 1972. *Professional Education Competency Needs of Vocational-Technical Programs in Post-Secondary Schools.* (Unpublished doctoral dissertation, Kansas State University.)

GILLIE, A. C. 1973. *Principles of Post-Secondary Vocational Education.* Colombus, Ohio, Charles E. Merrill Publishing Co.

HUNSICKER, H. N. 1979. *Information Prepared for the 1979 Annual Report Relating to Vocational Education for Agriculture and Agribusiness Occupations.* (Unpublished manuscript.) [Available from: U.S. Office of Education, Washington, D.C.]

LOMBARDI, J. 1973. *Managing Finances in Community Colleges.* San Francisco, Calif., Jossey-Bass, Inc., Pubs.

MEDSKER, L. L. 1971. Strategies for Evaluation of Post-Secondary Occupational Programmes. *The Second Annual Post-Secondary Occupational Education,* Vol. 2, 1971, pp. 7-28.

MOKMA, A. 1979. *Directory of Two-Year Post-Secondary Programs in Agriculture, Agribusiness and Renewable Natural Resources Occupations.* Washington, D.C., U.S. Office of Education.

MONROE, C. R. 1972. *Profile of the Community College.* San Francisco, Calif., Jossey-Bass, Inc., Pubs.

PHILLIPS, D. S.; Briggs, L. D. 1969. *Review and Synthesis of Research in Technical Education.* 2nd ed. Columbus, Ohio, the Ohio State University.

RIENDEAU, A. J. 1975. Trends in Post-Secondary Vocational and Technical Education. In: M. E. Strong (ed.), *Developing the Nation's Work Force.* Washington, D.C., American Vocational Association.

SHERMAN, G. S.; PRATT, A. L. 1971. *Agriculture and Natural Resources Post-Secondary Programs.* Washington, D.C., American Association of Junior Colleges (ERIC Document Reproduction Service No. ED 054 362).

SIDNEY, H. 1968. *Methods of Teaching Agricultural Occupations in Community Colleges and Area Vocational Schools.* Washington, D.C., U.S. Office of Education (ERIC Document Reproduction No. ED 026 529).

——. 1978. The Emerging Partner in Agricultural Education—Post Secondary. In: P.R. Smith (ed.), *Agriculture Teachers Directory.* Saltsburg, Pa., Smith Publications.

Transitions in Agricultural Education Focusing on Agribusiness and Natural Resources Occupations. 1971. Washington, D.C., American Vocational Association.

SVOB, M. J. 1973. A Pilot Project for High School/Community College Articulation. *North Central Association Quarterly,* Vol. 47, 1973, pp. 281–5.

WHINFIELD, R. W.; JOHNSON, M. E. 1974. Post Secondary Expansion as a Part of Professional Development. In: K. Greenwood and J. D. Skinkle (eds.), *Advocacy on Issues.* St. Paul, Minn., University of Minnesota.

WILLIAMS, D. L. 1972. Post Secondary Technical Education in Agriculture. *Agricultural Education,* April 1972, pp. 261–2.

University education in agriculture

Glenn Z. Stevens and David L. Howell

There are agricultural programmes in seventy-six of the member institutions of the National Association of State Universities and Land-Grant Colleges (NASULGC). (See Tables 7 and 8.) Fifty-one are in institutions where the programmes were established by state legislative bodies acting under authority of the Morrill Act of 1862. Each of the fifty states has a programme in one institution; Massachusetts in two. Connecticut also has a second Agricultural Experiment Station not part of a college or university.

Sixteen states with programmes in agriculture are in land-grant colleges or universities established by legislatures under authority of the second Morrill Act (1890). In Alabama, Tuskegee Institute is a special member of NASULGC. The Act of 1890 provided funds for land-grant colleges and universities for black students. These institutions continue to serve minority populations, but today they also serve the rest of the population.

The more recent programmes established by special acts of Congress were the land-grant programmes in the University of the District of Columbia, University of Guam, University of Puerto Rico and College of the Virgin Islands. Their involvements in agriculture are structured to meet their specialized needs.

Three other large public universities with significant programmes in agriculture are members of NASULGC. They are Arizona State University, Southern Illinois University and Texas Tech University. These universities do have agricultural colleges, but they are not land-grant universities.

University programmes for the preparation of teachers of agriculture

In most states, the land-grant university, where the state college of agriculture is located, has been designated by the state board for vocational education to receive funds for support of an approved programme for preparation of teachers of agriculture. The functions to be performed, standards for training offered and qualifications of teacher educators are specified in the state plan for vocational and technical education. Occupational competency and professional skill in teaching are required of all persons who are issued certificates or credentials for teaching in schools receiving state and Federal funds. In agriculture it has been assumed that secondary-school graduates with farm backgrounds, especially if they have had secondary-school instruction in agriculture and employment in agricultural positions in the summers between college terms, are occupationally experienced. They primarily need agricultural science and professional education courses to qualify as beginning teachers. Students without agricultural work experience are required to obtain such experience before the completion of their degree. The supervised intern programme with the student working

TABLE 7. Fall enrolment of undergraduates in agriculture by classes and by year

Classes	1950	1955	1960	1965	1976	1977
Freshman	10 196	11 207	10 336	13 101	23 729	23 456
Sophomore	9 444	8 617	7 649	9 921	21 553	21 008
Junior	9 729	7 558	7 489	9 097	26 019	25 573
Senior	10 606	6 829	7 054	8 557	24 772	26 589
Other	1 375	846	891	1 081	1 655	1 893
TOTAL	41 350	35 057	33 419	41 757	97 728	98 519[1]

1. The enrolment for 1978 shows a decline from the 1977 total.

Source : Report to the Resident Instruction Section, Division of Agriculture, NASULGC, 1977.

TABLE 8. Resident undergraduate enrolment in agricultural and related curricula administered in the NASULGC member institutions, fall term, baccalaureate programme

States and institutions[1]	1963	1974	1977
Alabama, Auburn U.	474	1 049	1 436
Alabama A. & M.	61	192	264
Alaska, U. of	—	—	—
Arizona, U. of	496	1 092	1 612
Arizona St. U.[2]	247	412	481
Arkansas, U. of	272	526	829
Arkansas, Pine Bluff	70	73	120
California, U. of			
Berkeley	874	1 012	1 080
Davis	—	2 928	3 831
Riverside	—	1 276	1 389
Colorado St. U.	610	1 046	1 347
Conn., U. of	242	995	1 331
Delaware St. C.	206	743	975
Florida, U. of	222	943	1 190
Florida A. & M.	49	110	188
Georgia, U. of	726	988	1 471
Georgia, Fort Valley St.	38	118	165
Hawaii, U. of	102	330	465
Idaho, U. of	276	369	489
Illinois, U. of	961	1 693	1 815
Illinois Southern U.	607	832	1 172
Indiana, Purdue U.	1 471	3 185	3 732
Regional C.	—	463	331
Iowa St. U.	1 853	3 058	3 770
Kansas St. U.	720	1 918	2 344
Kentucky, U. of	401	1 221	1 412
Louisiana St.	607	788	813
Louisiana U. & A. & M.	60	135	184
Maine, U. of	498	1 445	1 880
Maryland, U. of	435	1 193	1 611
Eastern Shore	33	61	41
Massachusetts, U. of	345	2 315	2 756

University education in agriculture

TABLE 8.—*cont.*

States and institutions[1]	1963	1974	1977
Michigan St.	1 177	2 604	3 296
Minnesota U.	1 200	2 056	2 337
Mississippi St.	501	1 079	1 187
Alcorn St.	50	203	170
Missouri, U. of[3]	969	2 398	2 859
Lincoln U.	31	38	63
Montana St.	407	918	1 057
Nebraska U.	765	1 421	1 800
Nevada U.	115	453	480
New Hampshire, U. of	214	1 537	1 624
New Jersey, Rutgers U.	431	2 097	1 744
New Mexico St.	414	1 042	1 258
New York, Cornell	1 909	2 826	2 976
North Carolina St.	686	2 570	2 617
North Carolina A. & T.	467	135	206
North Dakota St.	561	763	1 105
Ohio St.	1 740	3 430	3 824
Oklahoma St.	1 168	1 734	2 103
Oklahoma, Langston U.	22	20	17
Oregon St.	673	1 136	1 216
Pennsylvania St. U.[4]	1 001	—	—
University Park	—	1 926	2 523
Commonwealth Campuses	—	1 032	1 270
Puerto Rico U.	161	499	812
Rhode Island U.	230	781	975
South Carolina, Clemson A.C.	555	602	897
South Carolina St.	34	5	—
South Dakota St.	759	1 297	1 366
Tennessee U., Knoxville	538	1 191	1 670
Martin C.	—	483	597
Tennessee St.	100	86	153
Texas A. & M.	1 027	3 185	4 275
Texas Tech.	937	1 406	1 505
Texas, Prairie View A. & M.	90	96	94
Utah St.	267	443	447
Vermont U.	212	911	787
Virginia Polytech.	460	1 976	2 456
Virginia St. C.	90	70	81
Washington St.	515	1 290	1 586
West Virginia U.	497	1 107	1 375
Wisconsin U.	741	1 877	2 511
Wyoming U.	266	422	508
TOTAL	34 952	81 736	98 519[5]

1. Students in veterinary medicine and home economics in these institutions are not included.
2. The majority of freshmen and sophomores do not transfer to upper-division colleges until completion of a two-year university college general education programme.
3. Includes non-traditional students.
4. Administered under one dean.
5. The enrolment for 1978 shows a decline from the 1977 figure.

Source : Report to the Resident Instruction Section, Division of Agriculture, NASULGC, 1977.

in an agricultural occupation for an extended period of time is one way of obtaining the needed experience. Academic credit may be obtained by the student if the work experience programme complies with the university standards and is under the direction of a university professor.

The general and liberal education of the bachelor's degree experience at a university contributes greatly to the preparation of agriculture teachers in development of leadership qualities and in the ability to inspire leadership in students.

Kellogg and Knapp (1966, pp. 92–3) surveyed the leading agricultural colleges and found that they 'are now placing more emphasis on education for long-term intellectual growth and less on how-to-do-it training in techniques for the first job'. Four main trends are: (a) many colleges have increased general education requirements; (b) there is a reduction in the number of technician-training courses in agriculture; (c) there are fewer tightly prescribed specialized curricula; and (d) more emphasis is placed on flexibility so that a student with the help of his counsellor can work out a suitable individualized programme. These changes have important implications for students preparing to teach agriculture in post-secondary schools and departments. Academic excellence encouraged by studying in greater depth in a particular field prepares and helps a young man to be more effective in his area of occupational specialization when employed on a faculty in a school of a size sufficient to offer varied programmes.

The one most valuable pre-service professional course is student teaching. The university teacher-education departments arrange for each student to spend ten to sixteen weeks working in an outstanding school in the state. The experience is an internship or may be likened to co-operative work experience, usually without salary or stipend. The trainee is accepted in the school as though he were a regular staff member, and he is given an opportunity to experience a cross-section of the normal teaching responsibilities and other duties of a teacher of agriculture. Co-operating teachers receive special workshop training for their supervision of student teachers. Teacher educators visit the student teachers and co-operating teachers several times to co-ordinate the learning experiences provided.

An example of the baccalaureate-degree requirements for a major in agricultural education is shown in Table 9. The requirements for the major allow for limited specialization in an area of agriculture. Many universities are now encouraging students to major in an area such as animal science and include the education classes as the electives. With more multiple agriculture teacher departments in secondary schools, specialization is very desirable.

The teacher education department generally accepts responsibility for the in-service education of employed teachers. This is done in co-operation with the state administrative staff and with the local school officials who supervise vocational teachers. Special attention is given by many universities to first-year teachers; individual visits and regional group meetings are scheduled regularly in some states.

There are financial incentives in most school-policy statements that encourage experienced teachers to enrol in off-campus courses for graduate credit. Teachers return to the university for Saturday, evening and summer courses to earn a graduate degree and permanent certification. It is not uncommon for one-third to one-half of the teachers in a state to hold a master's degree. While earning the degree, a thesis research experience increases the teacher's competence, and may produce results worthy of use in other schools. Thus, the research function of the university is extended.

The degree requirements for a Master of Science or Master of Education in most universities include twenty-four to thirty-three semester credits in course work. Professional education classes that include classes specifically in agricultural education should total twelve to eighteen of the total credits of course work. It is also recommended

TABLE 9. Distribution of the total of 132 credits required for the baccalaureate degree in agricultural education at the Pennsylvania State University

I. *Baccalaureate degree requirements : 58 credits*
   Communications (12 credits)
      English 10, 20 (6)
      Speech communication 200 (3)
      Other communications (3)
   Quantification (6 credits)
   Natural science (24 credits)
      Biology 11, 12, 13 (9)
      Chemistry 11, 34 (6)
      Other natural sciences (9)
   Arts and humanities (6 credits)
   Social and behavioural sciences (6 credits)
      Psychology 3 (3)
      History 20 or 21 (3)
   Health sciences, physical education
   and physical recreation (4 credits)
      Health sciences or nutrition (1)
      Physical education (3)

II. *Requirements for the major : 65 credits*
   Prescribed courses (24 credits)
      Agricultural education 10v (2)
      Agricultural education 12v (3), 13v (1), 14v (8), 22v (2), 418v (2), 422v (1)
      Vocational education 1v (2)
      Educational psychology 14 (3)
      Supporting courses and related areas (41 credits)
      Animal science (6–15)
      Plant science (6–15)
      Business (6–15)
      Engineering (6–15)

III. *Electives : 9 credits*

Source : The Pennsylvania State University Bulletin, 1979, p. 56.

that six to twelve credits of course work be taken outside the field of education. These could be in the technical area of specialization such as animal science. The student is also encouraged to include several courses in the area of research in preparation for conducting a research project as a part of the degree programme or to develop an understanding of how to interpret the results of research for their use. A maximum period of six years is usually allowed for completion of the degree.

Many of the agricultural education programmes in the United States also offer Doctor of Philosophy (Ph.D.) degrees. In some universities, these programmes are for Ph.D.s both in agricultural education and agricultural extension. Approximately seventy-five semester credits of graduate-level course work beyond the baccalaureate degree and an acceptable thesis are required in most programmes. Approximately two-thirds of the total course work would be in the candidate's major field. A minimum of approximately fifteen credits would be completed in a minor field which contributes to the candidate's needs and interests.

The degree of Doctor of Education (D.Ed.) is also offered by many of these same agricultural education programmes. This degree is also available in many universities to

49

those in both agricultural education and agricultural extension. In most universities the requirements for the Ph.D. and D.Ed. are the same, although at one time the Ph.D. degree required a foreign language and the D.Ed. did not. Most universities have now dropped this requirement.

The courses to be included in the Ph.D. or D.Ed. programme are identified by the candidate with the approval of the doctoral committee. Emphasis is usually on course work in instructional processes, research and educational philosophy, with one or two minor areas indentified. The minor area could be international education, administration and supervision, counselling or curriculum development. A technical area of agriculture could be another possibility, e.g. animal science, horticulture or agricultural engineering. The candidates' interests and occupational goals would be a prime consideration in the selection of courses for the doctoral programme.

University programmes of research in agricultural education

It is necessary in the behavioural sciences as in the physical and biological sciences to recognize the complementary functions of research, development and dissemination. Basic research is not only conducted in laboratory situations; it is most likely to be carried out by well-trained research specialists in the field. Development in industrial research organizations follows basic discoveries. Schools, however, are structured so that some developmental projects are organized to compare techniques and procedures despite the lack of basic research antecedents. There is the other complaint that the time lag between productive basic research and effective general dissemination is too long. State directors, as administrators of research programmes and funds, must give these issues serious consideration. Project proposals designed by university research staff members must be cognizant of competing objectives.

A Research Coordinating Unit (RCU) was established in each state with funds of Section 4-c of the Vocational Education Act of 1963. The RCU staff members assist universities and schools through their research and development projects in vocational education.

In agricultural education there have been regional and national research co-ordination efforts since the 1920s. A continuous series of abstracts of studies has been published since 1935. The *Summaries of Research and Development Activities in Agricultural Education in the United States of America* contained abstracts of 140 selected research studies. Thirty-three were staff studies, fifty-one were doctoral dissertations and fifty-six were masters' theses. Some topics included were agricultural engineering, attitudes, developing nations, evaluation, teacher education and vocational agriculture teachers. In recent studies, there has been a steady advance from simple description to sampling survey and experimental research in the research designs. The use of analytical sampling techniques may be comparative-causal in nature, and they usually involve controls, randomization and replication.

Agriculture provided the classic, and first, examples of experimental research in which analysis of variance models was used. Research in education uses covariance analysis in varied experimental designs. Applications of computers to school needs is becoming increasingly important. Research may now be designed which would have been prohibitively laborious only a few years ago. But it must not be forgotten that the 'software', the ideas, the decision-making ability of men, rather than the 'hardware', the electronic equipment, will determine the ultimate benefits obtainable from effort applied to improve research designs.

The publication *You and Utilization Research* (United States . . ., 1963, p. 3) concluded its appeal to all vocational and technical educators to engage in research with this encouragement for co-operation:

If one fact stands out above all others, it is that the problems of research cannot be solved by any one agency. The cooperative efforts of all groups concerned with vocational and practical arts education and its products must be focused on the problem. Labor and management groups, farm organizations, business and distributive enterprises, and homemakers' groups should be called upon for assistance. Many such groups, or individual firms, are interested in solving specific problems and are willing to work with the schools on cooperative research studies. School people must team up with other groups to find solutions which are quickly applicable so that the vocational and practical arts education of tomorrow will be the best we can conceive today.

Evaluation of local programmes
of vocational-technical education in agriculture

Evaluation of an educational programme is an appraisal of outcomes in terms of objectives with due consideration of the ways and means employed to attain worthwhile goals. The chairman of the AVA national committee carefully differentiated programme objectives of a school, a state or the nation from educational objectives appropriate for an individual student. Teachers and schools should engage in continuous self-evaluation of programme objectives. State and national evaluations tend to be periodic and to rely less upon first-hand observation and more on analysis of data in formal written reports.

Evaluation in agricultural education may be traced to the early years of Smith-Hughes vocational programmes. The product of the schools was studied in terms of job placement and advancement. It had been assumed that meeting the labour requirements of industry was the objective of trade and industrial education and, therefore, agricultural education should be judged on the same basis. Some years later, the ways and means approach that had been developed for evaluation of secondary schools was adapted to agriculture by a national committee and recommended to the states. The relative merits of evaluation of the process or of the product of vocational and technical education have been argued in many professional meetings and journal articles. Hensel (*Education and Programme Planning* . . ., 1966, pp. 9–11) keynoted a national evaluation seminar by urging that both be considered and that criteria be developed recognizing each viewpoint. He cautioned against the possible frustrations of barriers to open and objective evaluation such as a supersensitivity to criticism, the inertia of tradition, absence of positive rewards for superiority, threatened security of personnel and uncertainty as to how to proceed.

Principles of evaluation have emerged over the years. Because the unit of school control is the local institution, usually administered by a school district and serving the people of a community, it was from this standpoint that Sutherland (1966, pp.13–18) recommended the following:

Evaluation of educational programmes should be made in terms of the objectives of these programmes.

Evaluation should include assessments and appraisals of both product and process.

Evaluation should be a continuous process, not just a 'point in time' judgement.

Evaluations should be made by teams composed of both professional and lay persons.

Evaluations of publicly supported programmes should include economic factors and be concerned with input-output relationships.

Evaluations and appraisals should be made not only on the basis of what has been done, but also on what has not been done.

The major purpose of evaluation should be to provide quality control and a basis for intelligent change.

Evaluations should be concerned primarily, if not exclusively, with the key indicators of success or failure.

Relevance for the human aspirations of people who elect to commit their productive work careers to the field of agriculture should guide teachers in this broad field as they expend their creative talents and emotional energies to promote the freedom of each student to achieve his maximum potential within a sound philosophical framework. Work is a fundamental element in the development and integration of personality. It has social significance of unlimited proportions. Superior and average students readily obtain and profit from professional and technical education. For the disadvantaged, vocational-technical training is a liberating experience, lifting the person from unemployment and impoverishment to independence and human dignity. Occupational education is a right and a necessity as well as an opportunity of unbounded dimensions. It is one of the best investments that can be made by an individual, community, state and nation. Evaluations to ascertain a programme's effectiveness and contribution to the lives of vocational students is and must be a continuous process by those in education.

Bibliography

A Report to the Resident Instruction Section. 1977. Washington, D.C., Division of Agriculture, National Association of State Universities and Land-Grant Colleges (NASULGC).

Evaluation and Program Planning in Agricultural Education. A Report of a National Seminar. 1966. The Center for Vocational and Technical Education. Columbus, Ohio, The Ohio State University.

KELLOGG, C. E.; KNAPP, D. C. 1966. *The College of Agriculture: Science in the Public Service.* New York, N.Y., McGraw-Hill Book Co.

Objectives for Vocational and Technical Education in Agriculture. 1966. Washington, D.C., Office of Education, U.S. Government Printing Office. (Bulletin No. 4, 1966.)

Summaries of Research and Development Activities in Agricultural Education in the United States of America. 1978. Department of Higher, Technical and Adult Education, School of Education. Storrs, Conn., University of Connecticut, December 1978.

SUTHERLAND, S. S. 1966. Objectives and Evaluation in Vocational Agriculture. In: *Evaluation and Program Planning in Agricultural Education. A Report of a National Seminar.* Columbus, Ohio, The Ohio State University.

The Pennsylvania State University Bulletin, 1979–1980. Baccalaureate Degree Programs. University Park, Pa., The Pennsylvania State University, Vol. LXXIII, No. 4, March 1979.

UNITED STATES AGRICULTURAL RESEARCH SERVICE. Eastern Utilization Research and Development Division. 1963. *You and Utilization Research. Career Opportunities at Eastern Utilization Research and Development Division.* Washington, D.C.

The Cooperative Extension Service

William I. Lindley

The Cooperative Extension Service in the United States has become the single largest informal education system in the world. The purpose of this programme is to help people improve the quality of their lives, develop their problem-solving skills, become competent consumers, wisely develop natural resources and build better communities. The emphasis is on helping people help themselves (King, 1975). The growth and success of this organization is a result of an ongoing co-operative effort of local people, county governments, the state land-grant universities and the United States Department of Agriculture, who share in planning and financing extension programmes.

The Cooperative Extension Service is a Federal agency having a direct educational link with people in their local communities. Because of its national affiliation, statewide organization and local support, the system has great flexibility and the ability to adapt short-range priorities to the long-range needs of the public. The Cooperative Extension Service influences the lives of both adults and youth. Educational programmes reach people on farms, in rural non-farm areas and in urban centres.

Historical background

The United States began as an agricultural nation. Its agricultural potential lay in the fertile valleys and plains where hardy settlers established farms, ranches and plantations. Nearly all the early leaders were farm or plantation owners, and agriculture was given high priority.

Agricultural organizations and societies were developed to encourage an exchange of ideas so that information about successful farming practices could be disseminated without delay. The first such society was founded near Philadelphia in 1785 (True, 1928). Agricultural societies offered both social and technical benefits to farm families and the movement spread rapidly through the mid-1800s.

In 1819, New York established the first state board of agriculture, which was primarily responsible for initiating one of modern Extension's predecessors—the farmers' institute (Vines and Anderson, 1976). Farmers' institutes were usually one- to five-day community meetings which dealt with several agricultural problems and with subjects related to the home. Informed speakers addressed the participants and group discussions were organized to promote a continuing exchange of ideas.

In the early 1800s the agricultural societies and state boards served as focal points of agitation for a Federal Department of Agriculture. At the same time, there was constant lobbying to promote the establishment of state universities to teach agriculture and mechanical arts (Sanders, 1966).

Agricultural legislation

In 1862 there were three Federal legislative agricultural acts that greatly influenced the creation of the present-day co-operative extension programme in America.

Morrill Act

In 1862, President Abraham Lincoln signed the Morrill Act, which provided for at least one university-level institution in each state, 'where the leading object shall be, without excluding other scientific or classic studies, to teach such branches of learning as are related to agriculture and the mechanic arts'.

The 'land-grant' act was designed to provide an endowment for each university by designating a number of 30,000-acre parcels of land to each state. The number of land parcels was equivalent to the number in each state's Federal Congressional delegation. The land was sold and the proceeds were used to purchase a college site, including an experimental farm. The remaining amount of money was invested to provide interest income for operating expenses.

Organic Act

There was more to come: the Organic Act created the United States Department of Agriculture (USDA) and the Homestead Act made millions of acres available to the public at almost no cost. The Organic Act provided the basis for widespread research activities, compilation of agricultural statistics and seed dissemination. The first responsibility of the newly established USDA was one of education. USDA was designed to gather and disseminate useful information on agricultural subjects. The cooperative extension service serves as the informal educational vehicle which takes the information to the general public.

Homestead Act

The Homestead Act, while providing an incentive for the westward migration of thousands of settlers, greatly reduced the value of the land-grants which were to provide the capital for agricultural study at the university level.

United States Senator Justin Morrill of Vermont immediately began a new campaign for additional funds to aid land-grant institutions. However, it was not until 1890 that another act was passed to increase support for the study of agriculture at institutions of higher learning. It was significant that this act required that land-grant institutions be open to both black and white students or that 'separate, but equal' institutions be established in states where integrated facilities were not available.

Hatch Act

University farms were used as agricultural teaching laboratories and the idea of research as a basis for course content gained widespread support. Federal government experiment stations were established as a result of the Hatch Act which was passed in 1887. This act provided for the establishment of an agricultural experiment station in connection with the land-grant university in each state.

Smith-Lever Act

The dissemination of information to the general public at this time was still being done through the farm organizations and societies. The idea of extending knowledge in an informal manner to the general public grew and matured. Extension-type work done by Kenyon L. Butterfield, President of Massachusetts Agricultural College, and Seaman A. Knapp of USDA had a strong bearing on the congressional decision to establish a

Cooperative Extension Service. Knapp pioneered the idea of using local farm demonstrations as a means of teaching improved agricultural practices and pest control. Butterfield urged a combination of farm demonstrations and a state university extension programme aided by Federal funding to carry information to farmers and the general public. The efforts and accomplishments of Butterfield and Knapp led to support of extension-type work. Extension agents were hired with local funds as early as 1906, eight years before Federal legislation called for such action in the Smith-Lever Act. It was this act in 1914 that established the Cooperative Extension Service as a part of the land-grant institutions and made it an equal partner with the experiment station and resident education.

Significant features of the Smith-Lever Act of 1914 included:

Extension work in agriculture and home economics was to be carried on by a land-grant university in co-operation with USDA. Local participation was required by this act in developing and carrying out the plans of work in each state.

Extension workers were to aid in diffusing useful and practical information related to agriculture and home economics to the people of the United States and to encourage application of this information. The audience was to be anyone not in residence or attending a land-grant institution.

Ten thousand dollars were to be given to each state by the Federal Government to support this effort during the first year. Additional amounts were to be pro-rated according to population in following years.

Amounts above the basic $10,000 grant were available only when matched by state or local funds. This provision was very important in achieving local participation in the extension programme. This was not a government handout; it encouraged local support and leadership activity.

The only control over the programme established by each university was to be an annual audit of funds by the Secretary of Agriculture. This was to ensure that the funds were spent within the purposes of the act. The act provided for extension agents of the land-grant university to be assigned to each county. Agents were to provide technical, social and economic leadership to the rural population. In addition agents were to be the eyes, ears and voice of the agricultural faculty and the experiment station in each county of every state in the union. The act not only provided for assistance to the farmer, but also for the rest of the family. Furthermore, it identified the need to work with youth through agricultural and homemaking clubs.

One section of the Smith-Lever Act that has helped to make the Cooperative Extension Service so successful was the provision for local participation. It is through such participation that extension programmes have developed strong community support. The executive committee is the legally-recognized county body with which the university, through its extension agents, co-operates in terms of programme development and support. Local people assist in planning the type of educational programmes they need. They are actually involved in helping the state universities identify priorities in the areas where service, new research and continued programming are needed. Local clientele attend programmes which they perceive as useful. Favourable responses are ensured when programmes are planned on the community level with local input.

In good times and bad, the Cooperative Extension Service has continued to serve the American public. Two World Wars, the depression of the 1930s, drought and other difficult times in history have created special needs which have stimulated extension workers to make each succeeding year more successful than the previous. The development of hybrid seed, artificial insemination of livestock, advances in the use of agricultural chemicals and new cultural practices have all presented challenges to extension personnel. The adoption process is based on an educational programme and extension personnel are agents of change. They are professional people who attempt to influence adoption decisions in the direction they feel desirable.

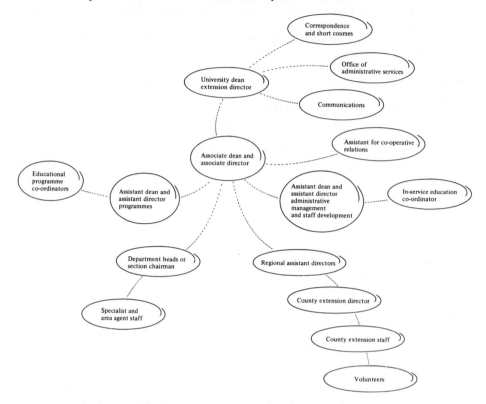

FIG. 3. Organizational chart, state-level Cooperative Extension Service.

The Cooperative Extension Service today

The Smith-Lever Act of 1914 has been amended and rewritten, and extension personnel still work to help people help themselves. The partnership of Federal, State and County agencies has served the public well even though legislative changes have altered the original course away from a primarily rural audience. Less than 5 per cent of the American work force is now engaged in production farming (Mayes, 1978). Population trends and the needs of a changing society have been addressed by an extension programme responsive to the needs of the general public.

Nutrition, urban gardening and community development programmes have taken extension programming beyond the farm boundary. This has led to a concerted effort to co-operate with a large number of agencies serving the general public. Societal demands are such that no one agency can serve the needs of all the people. This may be illustrated in the area of 4-H and youth work. 4-H, as the extension youth programme, is one of many youth-serving agencies. A spirit of co-operation and collaboration has led to bigger and better programmes serving young people with varied backgrounds. The 4-H motto 'To Make the Best Better' has been even more appropriate in recent years.

In the future the Cooperative Extension Service will serve the people in ways similar to those of the present and the past. It will continue to be responsive and flexible, as well as a full partner with university resident instruction and research.

General extension objectives

The objectives of educational programmes conducted by the Cooperative Extension Service are to improve the income-producing skills and the quality of life of people by providing educational assistance which aims to:

Produce farm and forest products efficiently while protecting and making wise use of natural resources.

Increase the effectiveness of marketing distribution systems.

Optimize development of individuals both as individuals and as members of the family and community.

Improve levels of living while achieving personal goals through wise resource management.

Improve communities through effective organization and delivery of services.

Develop informed leaders for identifying and solving community problems.

In striving to attain these objectives, Extension's basic principle is to provide people with knowledge and information on the alternatives to be considered when making sound decisions (Beattie, 1976).

Programme development

The triad of resident instruction, experiment-station research, and co-operative extension education operating out of each land-grant university has proved to be an unequalled combination of resources in bringing the most up-to-date methods and knowledge to bear on local problems. As part of the land-grant university, the Cooperative Extension Service draws from both the university's knowledge base and specialists working in academic departments.

The strengh of the Cooperative Extension Service is maintained through the involvement of clientele in helping to determine, plan and implement informal educational programmes that meet the needs of the local community. Extension philosophy is based on the belief that people must be reached where they are and at their level of interest and understanding (Beattie, 1976). This type of informal education serves people most effectively when they are involved in identifying needs in planning and programmes.

Programme development is an active, ongoing process which must be meaningful and understood by those involved. The programme planning process is based on past experience, present analysis and future implications. The programme development process consists of three phases: (a) planning, (b) implementation and (c) evaluation. As a continuous process, programme development may be illustrated in a circular pattern (see Fig. 4).

In summary, extension programmes must satisfy the people by: (a) being based on the local need, (b) presenting accurate information at a level of understanding suitable to the clientele, (c) being responsive to the ever-changing situation in each region or community, (d) continually evaluating programme offerings and related procedures, and (e) providing a variety of offerings that goes beyond the boundary of an agricultural advisory service. Effective extension programming has something to offer the entire family.

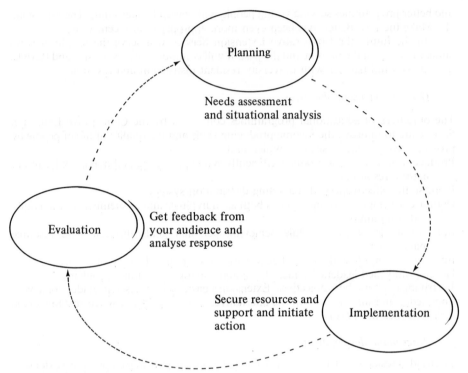

Fɪɢ. 4. Programme development.

Programme areas

Cooperative Extension Service education programmes may be broken into the four broad categories of: (a) agriculture and natural resources; (b) community development; (c) family living; and (d) 4-H and youth education. Each programme area may then be subdivided into related topics for programme planning purposes (see Table 10).

Agricultural and natural resources
The Cooperative Extension County Agricultural Agent is the farmer's link with agricultural research. The effectiveness of the extension education programme may be illustrated by the success of the American farmer. Research identifies and develops such things as hybrid seed and artificial insemination. It is the Cooperative Extension Service, through informal educational programmes, that transmits the information to the producer and assists him in implementing the new ideas.

On-farm demonstrations are used to show the results of new crop varieties and cultural practices. Field days and open house programmes at research stations are a constant source of knowledge and encouragement for agriculture producers. In nearly all of these activities, it is the extension agent who is the key figure in the education process and the adoption of new ideas.

There is an overall downtrend in the number of farms in the United States, but an increase in farm size. In 1960, an average American farm had 120 hectares; by 1977, the

TABLE 10. Main categories of the Cooperative Extension Service educational programmes

Agriculture and natural resources	Community development	Family living	4-H and youth education
Crop production	Community planning and land use	Human nutrition, food and health	Leadership, citizenship education and community development
Livestock production	Leadership	Family stability and human development	Animal, plant and soil sciences
Farm business management	Local government and finance	Family management of resources	Natural resources and environment
Agricultural farm marketing and farm supplies	Community services and facilities	Preventive health care	Home and family resources
Ecology, natural resources and environment	Housing	Improving housing and home environment	Health and safety
Mechanical science technology and engineering	Community economic development	Energy conservation	Creative and performing arts, leisure education and communications
Safety	Health	Cultural arts and heritage	Mechanical sciences and energy conservation

figure was 160 hectares. Larger farms mean higher individual investments for farmers. The Extension Agricultural Agent is active in helping farmers make decisions in the areas of profitability of enterprises, land use, equipment selection and finance. The vast majority of farms in the United States are owner-operated; some 89 per cent are operated by their owners or part-owners (Mayes, 1978).

Agriculture and Natural Resources priorities include: crop production; livestock production; farm business management; the development of a strong agricultural business community responsive to the needs of farmers, industry and consumers; mechanical science technology and engineering; continued efforts in the area of conservation of natural resources; and farm safety.

The agent may use a variety of teaching methods as he approaches people at the county level. Some of these techniques may include individual contacts such as farm and home visits, office calls, telephone calls, personal letters and result demonstrations. A second method is through the use of group contacts in which the agent may use demonstration meetings, conferences, tours and lectures. A third method, mass contacts, uses bulletins, newspapers, radio, television, exhibits and posters to reach large numbers of people. The Extension Agent must exercise good judgement in determining the most effective technique for reaching a particular audience. For example, the use of mass contact such as newspaper or radio would be appropriate in disseminating non-technical advice when speed is important. In contrast, a farm visit takes time and only one or two people are contacted with each visit. The individual contact, however, provides a good learning situation and allows the Extension Agent to gain firsthand knowledge of the farm and home conditions.

Community development

Community problems are a result of population changes, changes in living patterns, technological developments, changes in governmental and social structures, environmental concerns and the cost of providing facilities and services.

Local leaders are becoming increasingly aware of the educational role that the Cooperative Extension Service can play in the area of community development and the service itself is becoming more and more responsive as people ask for help in resolving community problems and achieving goals.

Its main target areas in community development lie within eight categories: (a) community planning and land use; (b) leadership development; (c) local government and finance; (d) community services and facilities; (e) housing; (f) community economic development; (g) health care and related problems; and (h) recreation and tourism.

Community development educational programmes are designed to meet the needs of lay people who find themselves in the position of being interested citizens or in the role of community leaders faced with the task of making decisions that affect the local population. Through educational campaigns, and with guidance and practice, community members learn to initiate and monitor ongoing activities. This process greatly increases the knowledge and assistance available at the local government level, particularly in rural areas and in small communities.

Community leaders and the general public need basic understanding of political, economic and social forces which influence public decision-making. There is an on-going need for educational programmes that address the fundamentals of meeting management, interpersonal communication and leadership in group discussion.

Extension Agents help citizens become aware of their community's economic potential by disseminating information and helping people to focus on activities that utilize or complement local resources and products. Many rural areas can benefit from a well-organized programme which promotes tourism. Planned programmes designed to bring increased numbers of visitors to a community can provide additional revenue, added employment opportunities and a higher sales volume of locally-produced products. Community development work results in better utilization of local resources. Educational programmes co-ordinated by extension workers must be supported by involved citizens who have had a role in identifying priorities and initiating a plan of action.

Family living

The basic mission of a family living programme is to help improve the quality of life for the family unit, individuals in the family and the family in the community.

Worldwide wage earners are faced with the problem of caring for their families with incomes that are diminished by inflation and rising costs of services. Family living problems place an emphasis on: (a) human nutrition, food and health; (b) family management of resources; (c) family stability and human development; (d) preventive health care; (e) improving housing and home environment; (f) energy conservation; (g) cultural arts and heritage; and (h) clothing and textiles.

Many people still suffer from malnutrition even though abundant, wholesome and nutritious foods are available at reasonable cost. The problem is one of education. Food and nutrition programmes are directed towards educating the consumer so that the food portion of the budget may be used more effectively. Several recent nationwide surveys have indicated that a significant percentage of Americans are consuming less than optimal quantities of iron, calcium and vitamins A and C—all key nutrients. Nearly one-third of Americans are overweight. With the percentage of females in the work force increasing at a rapid rate, about 30 per cent of the average American family food budget is used to purchase meals eaten outside the home in 'fast food' restaurants.

At the national level, the major focus deals with dietary goals and specific guidelines for improved nutritional standards. Co-operative extension nutrition programmes face the challenge of reaching every consumer with the message that good health and good eating habits are closely related (*Extension in the 80's*, 1979).

Rapid changes and rising expectations in an inflationary economy add stress to family management. Family management of resources has become one of the most serious problems that face the American family. Easy credit and high interest rates provide a particularly vexing problem for many families. Educational programmes with the emphasis on family resource management are provided on a regular basis. Extension Agents use a variety of methods to educate the consumer; these may include workshops, direct mailing and the use of radio and television.

Families develop their own values and establish their own goals; they create the environment in which children develop and adults relate to each other. Extension personnel work with families in the areas of child development and care, interpersonal communication skills and problem areas such as life-styles, values, status and family structure.

Access to adequate housing in a suitable location is a major problem facing American families. The average cost of a new home has increased to nearly $60,000. In addition, families are confronted with other housing problems such as: high interest rates, increased energy costs, higher taxes, costly repairs and an inadequate supply of new and rental housing. Extension programmes in this area are geared towards providing information and educational programmes which lead to greater satisfaction with the home environment and increased knowledge and ability to make wise housing decisions.

The energy crisis is real. The consumer is faced with high energy costs, and the United States is faced with the problem of being dependent upon other nations for liquid petroleum supplies. Cooperative Extension Service energy-related programmes lead consumers to a better understanding of the energy problem and the efficient use of non-renewable resources. Educational programmes, publications and the dissemination of factual information are some of the ways in which the family living Extension Agent can play a vital role.

Related arts programmes are designed to meet the needs of many age levels, interest groups and skills levels. By developing sensitivity to design principles, people will be able to make choices for long-lasting satisfaction in home environments, clothing, housing and other areas of daily life. The factors that raise the quality of life for people are pleasure satisfaction, and a sense of self-worth. Pleasure in achieving job satisfaction and the feeling that one has made a contribution to one's self and/or others can be achieved through related-art activities.

Family clothing and textile programmes help consumers make the best use of their budget allocations for clothing. Educational programmes are carried out in the areas of wardrobe planning, consumer education, textile selection and home sewing.

4-H and youth education

The American Cooperative Extension Service 4-H programme is known throughout the world. The green four-leaf clover with a white 'H' on each leaf represents the fourfold development of a young person's head, heart, hands and health.

In the early 1900s corn clubs for boys and canning clubs for girls were strictly rural in nature and, at least in part, aimed at teaching parents through their children. Early 4-H club work was clearly directed at improving farming skills for boys and homemaking skills for girls. By 1962 4-H clubs had over 2.33 million members. In 1978 nearly 5.5 million 4-H'ers made up the largest youth programme in the United States.

Today's 4-H projects and activities are designed to appeal to young people in rural, suburban and urban areas. Projects range from a dairy calf on a farm to arts and crafts in a city street camping situation. Interests of young people are met through such varied programmes as international youth exchange, community development, economics and careers.

4-H programmes offer young people the opportunity to develop leadership skills through club work, informal groups, camps, family programmes, special events and individual projects. Thousands of volunteer leaders, assisted by local Extension Agents, provide leadership for 4-H programmes. These volunteer leaders are a major factor in the success of 4-H; over the last ten years, their number has doubled. In 1978, over 578,000 volunteers contributed their time, talent and personal resources to support 4-H (Brown, 1979).

4-H is a voluntary educational programme for boys and girls and membership is open to any boy or girl between 8 and 19 years of age. Young people in a neighbourhood or community elect their own officers, hold regular meetings and 'learn by doing' under the guidance of local 4-H leaders.

Extension's purpose in this area is to provide programmes for youth through the involvement of parents, other adults and volunteers who organize and conduct learning experiences in a community setting. The focus is on human interaction designed to develop skills, abilities and understanding in youth and adults as participating and influential members of their communities. The central aim is for youth to acquire a set of skills for perceiving and responding to life's significant events. 4-H priorities include: (a) the improvement of family, parent, youth and sibling relationships; (b) the involvement of youth in community development, decision-making and social responsibility; (c) the assistance of youth in vocational development and decision-making; (d) the enhancement of social, mental and physical health; (e) the involvement of youth in clarifying individual and social values; (f) the assistance of youth in developing life-styles appropriate to changing times and balance in worldwide resources.

The 4-H motto 'To Make the Best Better' reflects the attitude of national, state and local leadership. The combined effects of public and private support, co-operative relationships of university and government, volunteer leaders, state and local extension personnel, family involvement, sound educational concepts and a dynamic process of programme development point to future growth without parallel.

Educational requirements for Extension Agents

Most county-level Extension Agents are hired with a degree in either agriculture or home economics. The programme area of 4-H and youth is the exception: agents may have degrees in a related social science. At the master's degree level, university work is usually done in a combination of areas that may include technical subjects, adult education, teaching methods, communication arts, personnel management, supervision and programme development. The master's degree is becoming increasingly important for advancement, and in some states it is a requirement for initial employment.

The role of the specialist

The extension specialist, as shown in the state level organizational chart, is recognized as an expert and consultant. Usually located in a university academic department, his responsibilities are primarily those of acting as a resource for county-level workers. In addition, he actively disseminates information through the mass media and may reach other clientele through workshops at the county level. The specialist provides leadership for extension programme development and in-service education at state and/or regional levels. This person develops, reviews, interprets opportunities and relevant materials, suggests goals and objectives for programme development and co-operates in single and interdisciplinary programme development.

Professional improvement through in-service education

As opposed to pre-service preparation, in-service education refers to an ongoing programme of professional improvement throughout an employee's career. In-service education is essential for the efficient operation of an extension service. The purpose of such training is to (a) up-date and fill in gaps in an individual's pre-service preparation; (b) develop additional cognitive, affective and psychomotor abilities needed for the effective fulfilment of an individual's role as an agent of change; and (c) stimulate continuous professional growth and achievement.

In planning a comprehensive in-service education programme, the following items should be considered: (a) long- and short-range goals of the extension organization; (b) responsibilities and duties of individuals; (c) recommendations of field staff, specialists, faculty with extension responsibilities and administration; (d) performance reviews; and (e) needs as perceived by the employee. In-service programmes may be held at a university where laboratories, experimental farms and other supportive teaching resources are available. Statewide or national programmes provide opportunities to bring together large numbers of staff and to create an interchange which reflects a wide variety of ideas and concerns. Some in-service programming lends itself better to regional or district sessions and individualized instruction may also be appropriate in meeting some very specialized needs.

In a word, co-operative extension in-service programmes help employees to cope with change and respond to new situations. The ultimate aim is to promote quality and efficiency which will contribute to increased educational and organizational effectiveness.

Summary

The Cooperative Extension Service is more than an advisory service to farmers. By including the programme areas of community development, family living and 4-H, in addition to agriculture and natural resources, a very large segment of the general public is served. Increasingly, agricultural production is the first priority, but service to youth, family and community has given extension programmes the strength and balance needed for continued support at all levels. Involvement of local people in programme planning,

William I. Lindley

the use of volunteer leaders, collaboration with other organizations and community action programmes make the extension service responsive to constantly changing needs.

Informal educational programmes, conducted by Cooperative Extension Agents, bring the message to the people in their local communities. Information delivery systems include discussion meetings, result demonstrations, direct mail, newspapers, radio, television, field days and open house events.

The alliance of research, resident instruction and extension education, all operating from a land-grant university base and supported by local, state and Federal funding, has produced a most effective informal educational system. Extension education is responsive to community needs, influenced by state priorities and supportive of government policy.

Extension work is best done where the people and the problems are located. It is not the job for individuals who feel that education exempts them from hard work or getting their hands dirty. Extension work is for the dynamic individual who likes people and is prepared to help people help themselves. Extension programmes in all countries can be improved only by assigning a high priority to local, state and national problems, devoting more human and monetary resources to the problems and providing extension employees with the best possible pre-service and in-service education.

Bibliography

ANDERSON, G. L. 1976. *Land-Grant Universities and Their Continuing Challenge*. East Lansing, Mich., Michigan State University Press.

BEATTIE, J. M. 1976. *The Cooperative Extension Service: Its Mission in Pennsylvania*. University Park, Pa., College of Agriculture, The Pennsylvania State University.

BROWN, N. A. 1979. *Update on 4-H. A Report on 4-H's Impact in the U.S. prepared by the 4-H Committee on Long-Range Planning and Budgets*. East Lansing, Mich., Michigan State University.

BROWN, N. A. et al. 1976. *4-H In Century III, 4-H*—Youth Programs. East Lansing, Mich., Michigan State University.

Extension in the 80's. 1979. University Park, Pa., Cooperative Extension Service, College of Agriculture, The Pennsylvania State University (unpublished planning document).

KELSEY, L. D. ; HEARNE, C. C. 1955. *Cooperative Extension Work*. 2nd ed. Ithaca, N.Y., Comstock Publishing Associates.

KING, T. B. 1975. *Induction Training Manual*. University Park, Pa., Cooperative Extension Service, College of Agriculture, The Pennsylvania State University.

KNOX, A. B. et al. 1979. *Yearbook of Adult and Continuing Education: 1978–79*. 4th ed. Chicago, Ill., Marquis Who's Who, Inc.

MAYES, D. (ed.). 1978. *1978 Handbook of Agricultural Charts*. Washington, D.C., United States Department of Agriculture. (Agricultural Handbook No. 551.)

SANDERS, H. C. (ed.). 1966. *The Cooperative Extension Service*. Englewood Cliffs, N.J., Prentice-Hall, Inc.

TRUE, A. C. 1928. *A History of Agricultural Extension Work in the United States 1785–1923*. Washington, D.C., U.S. Government Printing Office. (U.S. Department of Agriculture, Miscellaneous publication No. 15.)

VINES, C. A.; ANDERSON, M. A. 1976. Heritage Horizons—Extension's Commitment to People. *Journal of Extension* (Madison, Wis.).

Curriculum development, instructional materials and instructional services

Raymond H. Morton

Teachers involved in the education of present and future agriculturalists need adequate instructional materials for their teaching, and the skills to use them (Jacks, 1967). Agricultural education leads the vocational field in the successful development of high-quality instructional materials. This success is largely due to the efforts of local agriculture teachers who contribute ideas for materials which are then developed by commercial and public agencies and disseminated to all teachers.

Agriculture teachers at secondary and post-secondary levels develop much of their own local curriculum materials. Although they may use materials and suggested curriculum outlines prepared by regional, state and Federal curriculum materials services, they generally reorganize prepared materials to initiate and supplement their own curriculum plan called a course of study. The course of study is based upon community needs and resources.

Thus, the local agriculture programme is the locus of curriculum and instructional materials development in the United States; the local agriculture teacher is usually the foremost curriculum technician. Although curriculum and instructional materials are developed by commercial companies and university curriculum materials services, this development builds on the stated needs of local agriculture teachers.

The purpose of this chapter is to (a) give the reader a brief description of the curriculum and instructional materials development process in use in many parts of the United States at secondary and post-secondary levels of education; (b) list and describe examples of instructional materials; and (c) list and describe examples of instructional services provided to educators.

The term 'local vocational agriculture programmes' will be used in this chapter to include all agriculture education programmes offered by secondary and post-secondary educational institutions in a community for the preparation and continuing education of citizens interested in agriculture.

The term 'curriculum materials services' will be used to refer to all organizations—owned by or affiliated to public education, operated for profit or non-profit—regularly engaged in the development, preparation and dissemination of curriculum materials. The term 'curriculum materials' will be used to refer to all materials used by an agriculture teacher or agricultural educator to plan and execute instruction in agriculture. 'Commercial publishers' refers to organizations, neither owned nor operated by public education institutions, engaging regularly in developing, preparing and disseminating curriculum materials. To illustrate these differences, Texas A and M University operates a curriculum materials service; Interstate Printers and Publishers, Inc., is a full-time commercial publisher of agricultural curriculum materials; Ford Motor Company, Tractor Division, is an agricultural machinery business that also functions as a commercial publisher of training manuals and product manuals.

A curriculum development model for vocational agriculture

Curriculum materials development in the United States often includes the following six activities: needs assessment, development, field-testing and evaluation, revision, dissemination and in-service education for teachers who use the materials. These activities occur in various contexts, including local vocational agriculture programmes; curriculum materials service centres; and commercial publishing firms.

Needs assessment

Needs assessment is the first activity in the development of curriculum and instructional materials. The rationale for the development of useful instructional materials and curriculum is based on the premise that a job can be analysed to identify specific skills or knowledge needed to perform that job. Further, the skills and knowledge identified through this process can be used as the basis to develop relevant educational experiences. The list of skills and knowledge identified through a process called *occupational analysis* is first written as task statements (*Conduct . . .*, 1978). Task statements define a worker's role, i.e. what he or she does on the job. These statements are translated into competencies, which are skills or knowledge that the teacher must teach to train workers for a specific job or cluster of related occupations. Competencies to be taught are used by the curriculum developer to form the basis for the organization of units of instruction for inclusion in a complete description of the local educational programme of vocational agriculture called the course of study (*Develop a Course . . .*, 1978). Competencies are further translated as written student performance objectives that specify who is going to perform the competency, when the competency is to be performed and the degree of proficiency with which the competency is to be performed by the learner. With this information, the curriculum developer can organize and assemble curriculum materials sequentially to allow the learner to increase progressively the number of competencies he can perform and to raise the performance level of those competencies previously acquired (*Develop Student . . .*, 1978).

To illustrate how this system succeeds in describing an occupation in enough detail to organize instruction, consider the occupation of artificial inseminator. One duty listed is handling cattle. Assume that six tasks have been identified, using occupational analysis, for the fulfilment of the handling cattle duty: detecting œstrus, determining the time to inseminate, identifying the symptoms of reproductive diseases, maintaining breeding records, moving cattle and operating the breeding chute. For example, the task 'detect œstrus' could be written as a performance objective: 'The student will be able to detect œstrus in a cow at the completion of the course.' This student performance objective clearly tells the curriculum developer what is to be taught to the student; appropriate materials are then developed and assembled by the curriculum developer or teacher. A written and publicly-stated student performance objective also serves as a standard for evaluating the effectiveness of the curriculum materials used to teach this competency.

Needs assessments, including occupational analyses, are completed at many levels of the educational system. National and state needs assessment studies have been instrumental in identifying: competencies for agricultural occupations common to most parts of the United States; standards and/or competencies needed to educate vocational agriculture teachers adequately; and the needs of agriculture teachers for professional and technical education, i.e. in-service education.

Local needs assessment studies are usually completed by agricultural teachers with the help of school officials and the local agricultural businesses and farms. One technique of assessing local needs, a community survey, is widely utilized by teachers to identify present and future occupations in agriculture within the community served by the agriculture programmes (Stinson, 1978). A community survey can help the agriculture teacher and the advisory committee develop long-range goals for the agriculture programme; these goals can also be used as evaluation criteria for determining the programme's effectiveness.

Community advisory committees composed of business, civic and school leaders help the vocational agriculture teacher conduct and evaluate the results of a community survey; their main function is to validate the competencies identified before curriculum materials are developed. National and State Advisory Councils develop long-range goals that are implemented as legislation and serve as guidelines for community vocational education programmes.

In summary, needs assessment is the first and most important activity of the multilevel curriculum model and is a continuous process. The results not only help the teacher design curricula, but permit revision of the agriculture curriculum as programme goals change. Needs assessment, to assist the curriculum developer identify competencies essential in occupations for which workers need to be trained, takes place at the Federal, state and local levels in the United States. The procedure for conducting community surveys and occupational analyses is often taught as part of the university teacher education curriculum. National and state curriculum materials services also conduct needs assessment studies before preparing instructional materials. The local teacher is expected to adapt or alter curriculum materials to meet the learning goals of the local programme. He uses community surveys, occupational analyses and labour market surveys to collect background data for the next phase—development.

Development

Until recently, the development of curriculum materials at Federal, state and local levels was organized only around subject-matter areas, e.g. animal production. However, as Curtis (1978) pointed out, curriculum materials were not designed around the competencies needed to perform a job, nor were they organized around the various abilities and interests of students. For example, it was assumed that curriculum materials designed for one course in animal production would be adequate for a large number of agriculture programmes. While it is true that there are many animal husbandry skills common to most occupations in the animal-production industry, there are many that are not. As the geographical distances between local programmes become very large, jobs and students differ. To carry this example one more step: castration of livestock is an approved husbandry practice common to most areas in the United States; techniques vary, however—a shepherd in Arizona may use a bloodless method involving a rubber band, while the beef grower in Ohio may surgically remove the testicles. Hence, agriculture teachers in these two states may teach the same skill differently.

Differences also exist with respect to student abilities and needs. Such differences are considered important to agriculture teachers in the United States. One typically finds gifted and handicapped students, and academic and terminal students, in the same class with one teacher, studying the same subject. Different curriculum materials and teaching methods are almost always needed to accommodate these differences.

Generalized instructional materials and curriculum must be designed with sufficient flexibility to enable the agriculture teacher to modify or supplement them with local resource materials. It was for these purposes that Curtis (1978) proposed a

67

Vocational Curriculum Development Model for Agriculture that attempts to include all necessary aspects of curriculum and instructional materials development (see Fig. 5). Although this model is not used by all curriculum developers, models of similar design are commonly employed in the United States.

For purposes of curriculum development, four levels of occupational competency are utilized: operational, skilled, technical and professional. To illustrate this hierarchy with an example from agricultural mechanics: an operator drives a tractor, a skilled employee repairs it, a technician builds it and a professional agricultural engineer designs it.

The four levels of occupational competencies help the designer of curriculum and instructional materials in three ways: (a) students with different abilities and needs can select an appropriate level of instruction; (b) learning experiences and instructional materials can be arranged so that simple competencies can be performed first, followed by competencies of increasing complexity; (c) the design of instructional materials to include operational, skilled, technical and professional levels encourages students to achieve higher levels of achievement.

The Curtis model also includes seven instructional areas in agriculture; here, no hierarchy is implied. The seven instructional areas (sometimes called agricultural taxonomies) provide the basis for organizing the subject-matter of agricultural education in the United States. Curriculum and instructional materials are classified using this

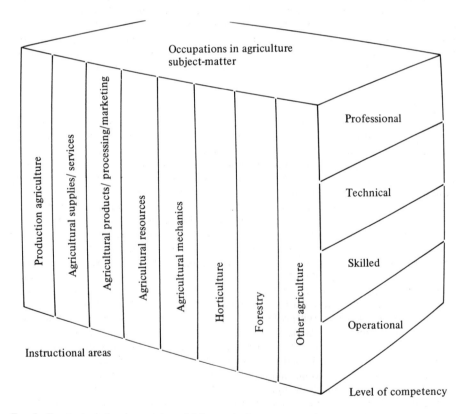

FIG. 5. Curriculum-development model for agriculture.

system. The second chapter of the present book contains a description of each taxonomy.

Occupational competencies needed for employment in agriculture are the contents of the cube. Competencies identified through occupational analyses previously described are placed at the appropriate level as the curriculum is being designed.

It is important to note that teaching students to perform a series of increasingly difficult competencies involves more than teaching a skill, e.g. sharpening a twist drill. Curriculum development for agricultural education moves beyond simple skill training to background knowledge and attitudes explaining why the operation is being performed. To take the previous example, the student is taught how the twist drill works, why it is important to keep it sharp and how to keep it sharp. Learning is increased because long-term retention is aided when students are taught why and how a competency is performed. The curriculum developer may therefore include, as part of the course of study, instruction in the theory, related concepts and attitudes needed to perform a cluster of skills required for worker performance and worker satisfaction.

Curriculum materials are also developed to include the opportunity to apply the knowledge and practise the skill needed to perform the competency in various learning environments: in the classroom, in the laboratory or school farm and on the job. For example, a lesson in agricultural mechanics might be developed to teach students how to adjust a mould-board plough for proper tillage, in this instance a laboratory experience. The curriculum developer would also be expected to include a teaching component allowing the teacher to measure and evaluate how well the student was able to make necessary plough adjustments for different soil types in the field and under minimum supervision.

Supervised occupational experience (SOE) programmes permit secondary and post-secondary students to apply the approved agricultural practices learned in the class and laboratory towards the production of an agricultural crop (plant or animal) for the student's own profit.

Instructional materials have been developed by curriculum materials services to instruct students in how to conduct an SOE programme. For adult students engaged in full-time agriculture, production guidelines and goals are developed by curriculum materials services with the help of agriculture teachers and extension agents for local situations. In this way, every farmer can compare his production to a standard and his own farming techniques to those of his neighbours.

Curriculum materials are also developed for YFA. (See the second chapter of the present book.) Curriculum materials have been developed by national and state curriculum materials services for use in secondary and post-secondary local vocational-agricultural programmes in the areas of leadership training, community development training and recognition programmes for youth and adult farmers. Each teacher uses these materials to teach rural youth and adult farmers how to organize and operate a local FFA or YFA group.

In summary, there are three important curriculum components in vocational agriculture at the secondary and post-secondary levels of instruction in the United States: the classroom and laboratory, supervised occupational experience, and vocational student organizations. For each area, curriculum and instructional materials are continually developed to be included in the local course of study for vocational agriculture. The teacher has freedom to choose how much, and in what way, national and state-developed curriculum materials are used. Each local vocational teacher is expected to develop a course of study that meets the needs of the local community and the needs and abilities of the students enrolled. No nationally-mandated curriculum is taught to all students, nor is there a required state or national examination for agricultural students at any educational level.

Field-testing

Field-testing or pilot-testing is the evaluation of curriculum materials that have been tried out, to determine their value, on a small sample of students. Field-testing is usually completed in classrooms, randomly selected to eliminate bias. It is supervised by agricultural educators with research skills and involves testing scientific hypotheses using statistical procedures.

Curriculum materials developed by individual teachers may also be field-tested in selected classes. At the community level, the vocational-agriculture teacher solicits the help of the advisory committee, school administrators and students as new curriculum materials are used initially. Generally, field-testing by the community agriculture teacher only involves collecting opinions; however, an increasing number of agriculture teachers are conducting controlled experiments to field-test curriculum materials. In-service education in research methodology is being provided to more teachers to help them with this process.

Only a fraction of the curriculum materials developed in the United States is ever field-tested; the process is expensive and time-consuming. However, it is acknowledged that proper field-testing of curriculum materials is essential if materials are to be utilized by others with the likelihood of their being useful in instruction.

Revision

Data collected from field-testing are analysed by the curriculum developer. Revision of curriculum is almost always necessary because (a) the competencies required to perform a job change as a result of new technology; (b) community and student needs change; (c) new processes are devised for production of instructional materials; and (d) national and state priorities shift and policies change.

Needs assessment is the most often used technique to collect data on which to base revision decisions. Follow-up studies of agriculture graduates provide information regarding the usefulness of curriculum materials. Periodic community surveys of businesses employing agriculture graduates provide information about changes in task statements.

Dissemination

Most of the instructional materials developed in the United States are distributed at minimum cost to the farmer, extension worker or teacher.

Within the last decade, state curriculum materials services have initiated co-operative arrangements to make the dissemination of materials more cost-efficient and widespread. Member states may join a regional consortium that supplies them with a catalogue listing all materials for sale, usually sold at cost. Curriculum materials are produced, stored and distributed to individuals in the member states under this arrangement. Not all curriculum materials services are operated as regional consortia.

The main advantage to the agriculture teacher of membership in a regional curriculum materials consortium is that he only needs to look in one catalogue to select and purchase most of his educational materials. The main disadvantage is that the consortium must charge the teacher more for items ordered if it has to carry a large and varied inventory.

Government agencies, such as USDA, disseminate thousands of different publications at minimum cost through the United States Government Printing Office

(Washington, D.C.). Instructional materials and services available for dissemination to agriculture teachers and farmers are listed below.

In-service

The last step in the Vocational Curriculum Model for Agriculture is the education of teachers or extension agents who use materials prepared by others for their own educational programmes. Teachers or agents who choose curriculum materials prepared by a curriculum materials service often receive in-service education from the disseminating organization on how to properly use and evaluate curriculum materials that they have selected. If the school participates in a curriculum materials consortium, in-service education is often provided. Commercial publishers selling curriculum materials to educators usually do not provide in-service education.

Instructional materials services

Instructional materials include: (a) all printed information, e.g. textbooks, reference books and pamphlets; (b) non-projected visuals, e.g. charts, posters and models; and (c) projected visuals, e.g. films, photographic slides, transparencies and video-tape recordings.

The primary functions of a curriculum materials service are to select, procure, evaluate and disseminate educational media for use in local agricultural programmes (Ridenour and Woodin, 1966). Funds to operate curriculum materials services emanate from the United States Office of Education, state departments of education, sponsoring universities, the receipts of sales of instructional materials and contributions from agribusiness and farmer organizations.

Curriculum materials services operate as a part of university teacher education departments or as separate curriculum materials laboratories. There are advantages to operating curriculum materials services as units within university teacher education departments because of: (a) interrelationships that ought to exist between teacher-education and instructional materials development; (b) accessibility and ease of co-ordination with subject matter specialists; (c) opportunities to interrelate research, development and evaluation through utilization of a 'critical mass' of talent; (d) facilitation of administrative procedures within a university setting; and, (e) greater freedom from political or state department pressure (McCracken, 1978).

Curriculum development services can help solve three problems teachers of agriculture could otherwise have by condensing the prodigious volume of new technical and professional knowledge into manageable form, finding time needed by the teacher to keep informed about new knowledge, and preparing and organizing materials structured in a logical sequence for instruction (McCracken, 1978).

Summary

Effective teaching by vocational agriculture teachers in the United States is largely achieved through adequate teacher education, both pre-service and in-service, and the effective use of high-quality materials in sufficient amounts, organized in a logical manner.

The teacher is the most important resource person for curriculum development in the United States. Agriculture teachers develop much of their own instructional material.

The course of study prepared by each local vocational agriculture programme is the curriculum plan used in planning, organizing, executing and evaluating instruction. It is based upon state and national priorities and pertinent analyses of agricultural occupations and student needs and abilities.

There are many systems for designing curriculum materials. Most utilize a system organized around the competencies required to perform the jobs identified as essential to the agricultural economy of the community and satisfy the needs of the students enrolled in the programme. Agricultural education curriculum, as distinct from occupational training curriculum, includes both components—the needs of students and the competencies for the job. A curriculum design model was discussed including both of these aspects.

The teacher can obtain help from several sources in preparing or organizing curriculum materials: (a) the local advisory committee; (b) curriculum materials prepared by curriculum materials services; (c) commercial publishing companies; and (d) other agricultural organizations.

The future of the agricultural economy in any nation is dependent upon an effective delivery system for new agricultural technology. There is a need to organize and transmit the rapidly increasing amount of technical knowledge required to produce agricultural goods and services for the world. Curriculum materials development is essential to achieve this goal.

Bibliography

Conduct an Occupational Analysis, Module A-7. 1978. Columbus, Ohio, The National Center for Vocational Education.

CURTIS, S. M. 1978. A Multi-Level Curriculum Model for Vocational Education in Agriculture. *Journal of the American Association of Teacher Educators in Agriculture.* Vol. 19, 1978, pp. 11–16.

Develop a Course of Study, Module A-8. 1978. Columbus, Ohio, The National Center for Vocational Education.

Develop Student Performance Objectives, Module B-2. 1978. Columbus, Ohio, The National Center for Vocational Education.

JACKS, L. P. 1967. *Development and Use of Subject Matter Materials for Vocational Education in Agriculture.* Jackson, Miss., State Department of Education, Division of Vocational and Technical Education.

McCRACKEN, J. D. 1978. Instructional Materials Development Units Should be Administered by Teacher Education Departments. *Journal of the American Association of Teacher Educators in Agriculture.* Vol. 19, 1978, p. 2.

New Instructional Materials for Agricultural Education. 1979. Athens, Ga., American Association Vocational Instructional Materials.

RIDENOUR, H. E.; WOODIN, R. J. 1966. *Guidelines for a State Vocational Agriculture Curriculum Materials Service.* Columbus, Ohio, The Ohio State University, Department of Agricultural Education.

STINSON, R. F. 1978. *Job Opportunity Based Program Planning for Vocational Agriculture; A Working Model.* University Park, Pa., The Pennsylvania State University, The Department of Agricultural Education.